D1346186

Viral Transformation and Endogenous Viruses

ACADEMIC PRESS RAPID MANUSCRIPT REPRODUCTION

1974 ABRAHAM FLEXNER SYMPOSIUM
VIRAL TRANSFORMATION AND ENDOGENOUS VIRUSES
HELD AT VANDERBILT UNIVERSITY, NASHVILLE, TENNESSEE
APRIL 1-2, 1974

174079

*Symposium on Virus Transformation
and Endogenous Viruses.*

Viral Transformation and Endogenous Viruses

EDITED BY

Albert S. Kaplan

Department of Microbiology
School of Medicine
Vanderbilt University
Nashville, Tennessee

QR472
S93
1974

ACADEMIC PRESS   New York   San Francisco   London   1974

A Subsidiary of Harcourt Brace Jovanovich, Publishers

COPYRIGHT © 1974, BY ACADEMIC PRESS, INC.
ALL RIGHTS RESERVED.
NO PART OF THIS PUBLICATION MAY BE REPRODUCED OR
TRANSMITTED IN ANY FORM OR BY ANY MEANS, ELECTRONIC
OR MECHANICAL, INCLUDING PHOTOCOPY, RECORDING, OR ANY
INFORMATION STORAGE AND RETRIEVAL SYSTEM, WITHOUT
PERMISSION IN WRITING FROM THE PUBLISHER.

ACADEMIC PRESS, INC.
111 Fifth Avenue, New York, New York 10003

United Kingdom Edition published by
ACADEMIC PRESS, INC. (LONDON) LTD.
24/28 Oval Road, London NW1

Library of Congress Cataloging in Publication Data

Symposium on Virus Transformation and Endogenous Viruses,
    Vanderbilt University, 1974.
    Viral transformation and endogenous viruses.

    Bibliography: p.
    1.   Cell transformation–Congresses.   2.   Oncogenic
viruses–Congresses.   I.   Kaplan, Albert S., Date
ed.   II.   Title.   [DNLM: 1   Cell transformation,
Neoplastic–Congresses.   2.   Oncogenic viruses–Congresses.   QZ202 V813 1974]
QR472.S93 1974   616.9'92'071   74-17148
ISBN 0–12–397060–1

PRINTED IN THE UNITED STATES OF AMERICA

Contents

List of Contributors

Boldface Denotes Speakers

MAX ARENS, Institute for Molecular Virology, St. Louis University School of Medicine, St. Louis, Missouri 63109

CLAUDIO BASILICO, Department of Pathology, New York University School of Medicine, New York, New York 10016

KAREN BEEMON, Department of Molecular Biology and Virus Laboratory, University of California, Berkeley, California 94720

PAUL BERG, Department of Biochemistry, Stanford University School of Medicine, Stanford, California 94305

S. BHADURI, Institute for Molecular Virology, St. Louis University School of Medicine, St. Louis, Missouri 63109

J. MICHAEL BISHOP, Department of Microbiology, University of California, School of Medicine, San Farancisco, California 94143

KARL BRACKMANN, Institute for Molecular Virology, St. Louis University School of Medicine, St. Louis, Missouri 63109

P. E. BRANTON, Department of Biology, Massachusetts Institute of Technology, Cambridge, Massachusetts 02139

STUART J. BURSTIN, Department of Pathology, New York University School of Medicine, New York, New York 10016

WERNER BÜTTNER, Institute for Molecular Virology, St. Louis University School of Medicine, St. Louis, Missouri 63109

JOHN CARBON, Department of Biochemistry, Stanford University School of Medicine, Stanford, California 94305

YOUNG C. CHEN, Department of Microbiology, University of Southern California School of Medicine, Los Angeles, California 90033

B. CORDELL-STEWART, Department of Microbiology, University of California School of Medicine, San Francisco, California 94143

PETER DUESBERG, Department of Molecular Biology and Virus Laboratory, University of California, Berkeley, California 94720

ROBERT R. FRIIS, Department of Microbiology, University of Southern California School of Medicine, Los Angeles, California 90033

KEI FUJINAGA, Institute for Molecular Virology, St. Louis University School of Medicine, St. Louis, Missouri 63109

DONALD J. FUJITA, Department of Microbiology, University of Southern California School of Medicine, Los Angeles, California 90033

*BETTY JEAN GAFFNEY, Stauffer Laboratory for Physical Chemistry, Stanford University, Stanford, California 94305

H. M. GOODMAN, Departments of Biochemistry & Biophysics, University of California School of Medicine, San Francisco, California 94143

MAURICE GREEN, Institute for Molecular Virology, St. Louis University School of Medicine, St. Louis, Missouri 63109

CARLOS B. HIRSCHBERG, Department of Biology, Massachusetts Institute of Technology, Cambridge, Massachusetts 02139

MICHAEL LAI, Department of Molecular Biology and Virus Laboratory, University of California, Berkeley, California 94720

HARRIET K. MEISS, Department of Pathology, New York University School of Medicine, New York, N. Y. 10016

*Present Address: Department of Chemistry, Massachusetts Institute of Technology, Cambridge, Massachusetts 02139

JANET E. MERTZ, Department of Biochemistry, Stanford University School of Medicine, Stanford, California 94305

FRED RAPP, Department of Microbiology, Milton S. Hershey Medical Center, Pennsylvania State University, Hershey, Pennsylvania 17033

PHILLIPS W. ROBBINS, Department of Biology, Massachusetts Institute of Technology, Cambridge, Massachusetts 02139

W. ROHDE, Departments of Microbiology and Biochemistry & Biophysics, University of California School of Medicine, San Francisco, California 94143

G. SHANMUGAM, Institute for Molecular Virology, St. Louis University School of Medicine, St. Louis, Missouri 63109

STEPHEN C. ST. JEOR, Department of Microbiology, Milton S. Hershey Medical Center, Pennsylvania State University, Hershey, Pennsylvania 17033

J. M. TAYLOR, Department of Microbiology and Biochemistry & Biophysics, University of California School of Medicine, San Francisco, California 94143

PETER TEGTMEYER, Department of Pharmacology, Case Western Reserve University, Cleveland, Ohio 44106

HOWARD M. TEMIN, Department of Oncology for Cancer Research, McArdle Laboratory, University of Wisconsin, Madison, Wisconsin 53706

DANIELA TONIOLO, Department of Pathology, New York University School of Medicine, New York, N. Y. 10016

PETER K. VOGT, Department of Microbiology, University of Southern California School of Medicine, Los Angeles, California 90033

G. G. WICKUS, Department of Biology, Massachusetts Institute of Technology, Cambridge, Massachusetts 02139

TADASHI YAMASHITA, Institute for Molecular Virology, St. Louis University School of Medicine, St. Louis, Missouri 63109

Preface

Through the generosity of Mr. Bernard Flexner an endowed lectureship was established in 1927 in the name of Dr. Abraham Flexner in recognition of his service to medical education and for his interest in Vanderbilt University. Each year an eminent physician or scientist is invited to serve as a lecturer in residence; Dr. George Klein of the Karolinska Institute, Stockholm, Sweeden was the Flexner lecturer for 1974.

It has been traditional to honor the Flexner lecturer at the end of his residence at Vanderbilt by holding a symposium which deals with the subject matter of the Flexner lectures. The symposium on "Viral Transformation and Endogenous Viruses" held at Vanderbilt University on April 1-2, 1974 continued this tradition. The papers presented at this symposium constitute the various chapters of this book.

When susceptible cells are infected with oncogenic viruses, a viral function(s) is expressed which sets off a chain of events leading ultimately to the transformation of the cells. The central theme of most studies on the oncogenic interaction between the virus and the host cell is a definition of the viral function(s) responsible for transformation. Most of the papers presented in this symposium dealt with this theme in studying either DNA- or RNA-containing oncogenic viruses. However, the type of changes which characterize transformed cells and the regulatory mechanisms which are altered after malignant transformation may also be studied using temperature-sensitive mutants of the host cells. Furthermore, changes in the cellular membrane after transformation are well-documented and the basis for these changes may contribute to an understanding of this phenomenon. The value of each of these avenues of approach is illustrated by some of the papers in the symposium.

The first part of the symposium was devoted to discussions of recent studies of the DNA-containing tumor viruses. The ability of Papovaviruses and Adenoviruses to transform cells has been known for some time and the main thrust of present research is towards an understanding of the molecular basis and the genetic controls for this phenomenon, issues which constituted a primary theme of this symposium. The papers presented dealt with the analyses at the molecular level of SV40 mutants with deletions, insertions, and duplications in their DNA, of the integration and transcription of adenovirus DNA, as well as the characteristics of temperature–sensitive mutants of these viruses.

Transformation by members of the herpes group of viruses is a latter-day observation. Herpesviruses usually interact lytically with their host cells. However, as indicated at the meeting, under the right conditions cells can be transformed by herpes simplex virus and cytomegalovirus, so that model herpesvirus-transformed cell systems are now available for study.

The second half of the symposium dealt with the chemistry and biology of the RNA-containing tumor viruses, viruses which are particularly useful reagents to study neoplastic transformation both in vivo and in vitro. A major aspect of these systems, which was discussed in this symposium, concerned the analysis at the chemical and genetic level of the genome of these viruses. These studies included also the partial characterization of a 4S RNA isolated from the 70S RNA of Rous sarcoma virus, which serves as a primer for the initiation of DNA synthesis by reverse transcriptase, as well as the characteristics of the RNA of recombinants of avian tumor viruses.

Recent evidence suggests that oncogenic viruses, particularly those containing RNA, may have evolved from normal cells. A considerable amount of work has been directed to the isolation and characterization of so-called endogenous viruses from normal cells. The last part of the symposium was devoted to biochemical and genetic analyses of endogenous viruses isolated from avian cells.

I should like to thank all those who contributed to the success of the symposium, particularly the following who served as session chairmen: Paul Berg, Fred Rapp, J. Michael Bishop, and Peter Vogt. I appreciate greatly the considerable help provided by Mrs. Marilyn A. Short, Administrative Associate, Division of Continuing Education, Vanderbilt University, in organizing the meeting and I am especially grateful to Mrs. Juanita Boyer, Office Manager, Department of Microbiology, whose cheerful efficiency eased my task enormously. Finally, I should like to thank the Virus Cancer Program, National Cancer Institute for its financial support.

Albert S. Kaplan

Vanderbilt University

Viral Transformation and Endogenous Viruses

ISOLATION AND CHARACTERIZATION OF INDIVIDUAL CLONES OF SIMIAN VIRUS 40 MUTANTS CONTAINING DELETIONS, INSERTIONS AND DUPLICATIONS IN THEIR DNA

Paul Berg, Janet E. Mertz and John Carbon

Department of Biochemistry, Stanford University
School of Medicine, Stanford, California

INTRODUCTION

Viruses bring new genes into the cells they infect. Generally the new genetic information serves to establish the machinery for multiplying the virus, specifically, the precursors and proteins needed to replicate and encapsidate the viral nucleic acids. With certain viruses and their appropriate hosts the infection has an alternative outcome. The infected cell survives and occasionally undergoes an heritable alteration of its morphology and growth characteristics, particularly, the ability to grow under conditions where normal cells are arrested and to initiate tumors in appropriate animals. Such transformed cells invariably contain and transmit at each cellular division all or part of the viral genome as an integral part of the cell's chromosomal DNA; moreover, expression of one or more of the integrated viral genes is required to maintain the transformed phenotype.

Each of these responses, viral multiplication and cellular transformation can be elicited with the small DNA virus, Simian Virus 40 (SV40). Because SV40 contains only a very limited amount of genetic information-5100 base pairs or enough coding capacity for about five proteins of 40,000 daltons molecular weight, one can be optimistic at the prospects of understanding the genetic control and molecular events of both the multiplication cycle and the transformation phenomenon.

1

What are these SV40 viral genetic elements and how are they organized physically on the viral DNA molecule? Furthermore, what are the viral gene products, how do they function and what is the arrangement of viral genes after integration into the cellular DNA? These are formidable questions and undoubtedly their answers will be some time in coming. It seems logical, however, to begin with the first question.

Several laboratories (1-5) have isolated and characterized conditional-lethal, temperature-sensitive (ts) mutants of SV40. Although such mutants can provide valuable information about the physiology and molecular biology of the viral life cycle (6-8), their utility is limited to genes that code for proteins and by the relative difficulty in accurately mapping the mutant loci on the viral DNA molecule. On the other hand substantial alterations in the viral DNA structure (e.g., deletions, duplications, insertions and substitutions) would very likely alter the normal viral phenotype and could be mapped by heteroduplex analysis (9, 10). This paper summarizes our general approach for the isolation, construction and propagation of such grossly defective SV40 mutants and presents a preliminary characterization of several representative defective mutants.

Coordinates for the SV40 DNA Map

Since the SV40 genome is contained in a continuously circular, that is, covalently closed double-stranded DNA molecule, one or more specifically located reference points are needed to fix the position of a particular genetic locus or region. Bacterial restriction endonucleases, enzymes which make double-strand cleavages in DNA molecules at specific nucleotide sequences, serve admirably to provide such coordinates (Fig. 1). A number of enzymes have proven to be quite useful for this purpose.

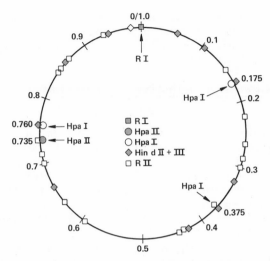

1.   Sites of cleavage of SV40 DNA by several restriction
endonucleases.

a) EcoRI (11) restriction endonuclease (12) cleaves
SV40 DNA once at the sequence$_{-CTTAAG-}^{-GAATTC-}$ to produce a
non-permuted population of unit length linear molecules
(9, 13, 14). We have defined the EcoRI restriction site as
position 0 of the SV40 DNA map (9) and the coordinates are
expressed as SV40 DNA fractional lengths.

b) Another useful enzyme, from Hemophilus parainfluenzae, is HpaII restriction endonuclease (15, 16). This
enzyme also produces only one double-strand scission per
SV40 DNA molecule at a position 0.735 SV40 fractional
length, clockwise, from the EcoRI cleavage site (0.735
SV40 map unit)(16, 17).

c) HpaI, a second enzyme from H. parainfluenzae,
cleaves SV40 DNA at 3 places and these have been mapped
at 0.175, 0.395 and 0.760 SV40 map unit (15, 16).

d) Several other restriction endonucleases cut SV40
DNA more extensively. A mixture of at least two enzymes
from Hemophilus influenzae (Hind II + III) cleaves SV40
DNA at eleven locations (shown on the map by diamonds (◆))
to yield eleven distinguishable fragments (18). Another
enzyme, EcoRII coded for by the RII plasmid in E. coli
(12), makes sixteen cleavages in SV40 DNA, one at each of
the squares(□)on the map (19). As illustrated below the

3

availability of these enzymes and a knowledge of their cleavage sites has permitted the selection, construction and characterization of SV40 mutants with alterations in the wild-type DNA structure.

Individual Clones of Defective, Helper-Dependent SV40 Virus

SV40 virus particles with defective genomes accumulate when plaque purified virus is serially passaged in susceptible cells at high multiplicities of infection (moi) (20-22). These defective viral DNAs are generally smaller than the DNA obtained from plaque purified virus and have gross alterations (deletions, substitutions, additions) in their molecular structure (21,22). Since the molecular defects contained in these DNAs are distributed somewhat randomly throughout the molecule (J. E. Mertz and P. Berg, manuscript in preparation), these preparations provide a rich source of mutants with substantial modifications in various portions of their genomes.

Our starting material for the isolation of such mutants was obtained by four serial passages of wild-type SV40 strain Rh911 (23)(WT800)·on primary African green monkey kidney cells using undiluted lysate obtained from each passage to initiate the subsequent infection. The SV40 DNA obtained from the fourth cycle of infection was judged to be grossly defective by the following criteria: the titer (plaque forming units (pfu)/ml) was more than two hundred-fold lower than that of an analagous lysate produced by low moi infection; the supercoiled DNA from the infected cells was very heterogeneous in size with molecules ranging in size from about 0.6 to somewhat greater than 1.0 SV40 fractional length; the defective DNA yielded deletion, substitution and addition loops when heteroduplexes with WT SV40 DNA were examined by electron microscopy; and, whereas about 99% of WT DNA preparations are cleaved once at map positions 0 and 0.735 by the EcoRI and HpaII restriction endonucleases, respectively, 55% of the DNA was resistant to cleavage by EcoRI endonuclease, 47% was resistant to HpaII endonuclease and 23% was resistant to both enzymes (J. E. Mertz and P. Berg, manuscript in preparation).

The EcoRI-resistant and HpaII-resistant DNAs were separated from the molecules which were cut by the enzymes by equilibrium centrifugation in cesium chloride-ethidium bromide gradients (24) and then used as the

4

sources from which cloned isolates of defective SV40 were recovered.

Preparations of EcoRI- and HpaII-resistant DNAs were four orders of magnitude less infectious than WT DNA in conventional plaque assays. Consequently, to isolate and propagate such defective genomes a procedure which permits complementation of their defect was required. Temperature-sensitive (ts) mutants of SV40 can be used for this purpose. To decide which of the known ts mutants would complement the growth of these defective DNAs, it was necessary to identify the defective function(s) in the EcoRI- and HpaII-resistant DNA populations. Following infection, the EcoRI- and HpaII-resistant DNAs induce the "early" SV40 specific antigens T and Y and replicate their DNA nearly normally, but they fail to produce the "late" virion capsid antigen. Consequently, molecules lacking either the EcoRI- or HpaII-restriction site (map position 0 and 0.735, respectively) can express those functions preceding and including DNA replication ("early" functions) but not the "late" functions; this result agrees with previous reports (25-27) that these two restriction sites are contained within the "late" region of the SV40 genome. On this basis tsA30, a non-leaky "early" mutant which can complement the "late" mutants tsB4 and tsB11 for growth (2) was selected as the "helper" virus to grow the restriction enzyme-resistant mutants. (These SV40 ts mutants were isolated and very generously sent to us by P. Tegtmeyer.)

To isolate individual clones of defective SV40 lacking the EcoRI and HpaII restriction sites, we coinfected monkey cells with the restriction endonuclease-resistant DNA preparations and tsA30 DNA, and then incubated the cells at 41°, the restrictive temperature for tsA30 growth. Although some of the plaques that appear were produced by a trace of WT molecules contaminating the defective DNA population or by viable defectives (see below), the majority of the plaques resulted from cells doubly infected with defective and complementing helper genomes. These putative mixed plaques were picked and plaque-purified by successive coinfections at 41° with added tsA30 helper virus. Generally, after one or two plaque-purifications, the virus suspension obtained from a mixed plaque produced 10-100-fold more plaques at 41° in the presence of added tsA30 virus than in the absence of the helper; and, whereas the former occurs with single-hit kinetics, the

5

latter more nearly approaches a two-hit process.

Eleven independent plaque isolates, 5 from the EcoRI-resistant DNA pool and 6 from the HpaII-resistant DNA pool, were selected for further study. DNA preparations and high titer virus stocks of these putative defective mutants were obtained by infection of monolayers of CV-1P cells with a mixture of tsA30 virus and virus from the last mixed plaque and incubation at 41°. At the appropriate time viral DNA or virus was isolated from the cultures by standard methods.

To obtain pure defective DNA from each of the clones, the mixed pool of closed circular DNA recovered from the mixed infection was incubated with EcoRI restriction endonuclease (if it was derived from the EcoRI-resistant DNA pool) or with HpaII restriction endonuclease (if it came from the pool of HpaII-resistant DNA) and the restriction endonuclease-resistant material from each isolate was then separated from the linear helper DNA by velocity sedimentation. For the eleven clones we have isolated the endonuclease-resistant or defective DNA represented between 50 to 96 percent of the total DNA recovered; with ten, however, more than 75% of the DNA was of the defective mutant.

Electron microscopic examination of the restriction endonuclease-resistant DNA fraction from each plaque isolate revealed essentially homogenous populations of circular molecules; the mean sizes of these mutant DNAs ranged from 0.71 to 0.96 SV40 fractional length. In a subsequent publication (J.E. Mertz and P. Berg, in preparation) we shall report the results of heteroduplex analysis of these defective DNA; those experiments show that the defective DNAs lack either the EcoRI- and/or HpaII-restriction endonuclease cleavage sites because they have extensive deletions of DNA encompassing these regions of the genome. Moreover, most of the mutants also have sizable duplications of the region of the SV40 genome that includes the origin of DNA replication (28, 29) which probably accounts for their preferential growth in mixed infection with the helper.

Individual Clones of Viable, Defective SV40 Virus

Each of the above cloned HpaII endonuclease-resistant mutants fail to produce plaques in the absence of helper

6

virus, but plaque-forming DNA molecules can also be isolated from the HpaII endonuclease-resistant pool of defective DNA which served as the starting material. Plaques produced from these are distinguishable from WT plaques by their smaller size, their late appearance and their slower rate of development.

After serial plaque-purification, ten individual viral clones repeatedly displayed the small plaque phenotype (Fig. 2); as anticipated, the viral DNA was completely resistant to cleavage by HpaII restriction endonuclease under conditions where more than 95% of the WT DNA is cleaved.

Is the resistance to HpaII endonuclease cleavage due to a base change in the restriction sequence or to a small deletion which removed the restriction site? This question was answered by comparing the electrophoretic mobility of mutant and WT fragments produced by Hind II + III cleavage of the respective DNAs. According to Danna and Nathans (18) the HpaII restriction sequence is contained within the Hind C fragment generated by Hind II + III cleavage at 0.65 and 0.76 map position (Fig. 3). Figure 4 illustrates the polyacrylamide gel electrophoresis fragment patterns obtained from a Hind II + III digest of a mixture of two representative ^{3}H-labeled mutant DNAs and ^{32}P-labeled WT DNA: In each case, the mutant DNA digest lacks the Hind C fragment and contains a new, faster migrating fragment. Since electrophoretic mobility is approximately proportional to molecular length (Fig. 5), we conclude that the altered Hind C fragments produced with the ten mutant DNAs are between 80 (the smallest deletion) to 190 (the largest deletion) base pairs shorter than the WT Hind C fragment. Since the Hind D fragment is unaffected in any of the mutants, the Hind site at 0.76 must be intact.

The larger deletions can also be detected by heteroduplex analysis in the electron microscope. Denaturation and renaturation of a mixture of EcoRI-cut mutant and WT DNAs would be expected to yield linear heteroduplexes containing a small deletion loop at the region of the HpaII restriction site (Fig. 6). This expectation was confirmed by electron microscopic analysis of heteroduplexes formed with the mutant having the largest deletion (Fig. 7); approximately forty percent of the linear duplexes contained a barely visible discontinuity at about 0.245 SV40 fractional length from the nearer end (Fig. 8).

7

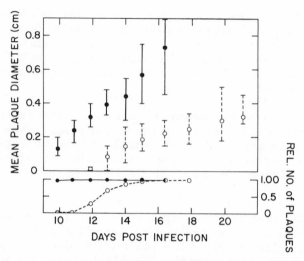

2. Size and rate of development of mutant and wild type virus plaques. Top graph: the mean plaque diameter (and range) of ten wild type (●) and mutant (o) plaques as a function of time after infection; Bottom graph: the number of wild type (●) and mutant (o) plaques at a particular time relative to the number of plaques observed sixteen days after infection.

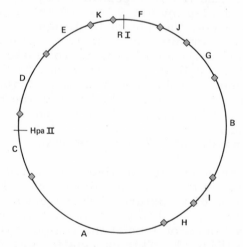

3. Sites of cleavage of SV40 DNA by EcoRI, HpaII and Hind II + III restriction endonucleases. The location of the Hind cleavage sites and the nomenclature for the fragments is taken from the work of Danna and Nathans (18).

8

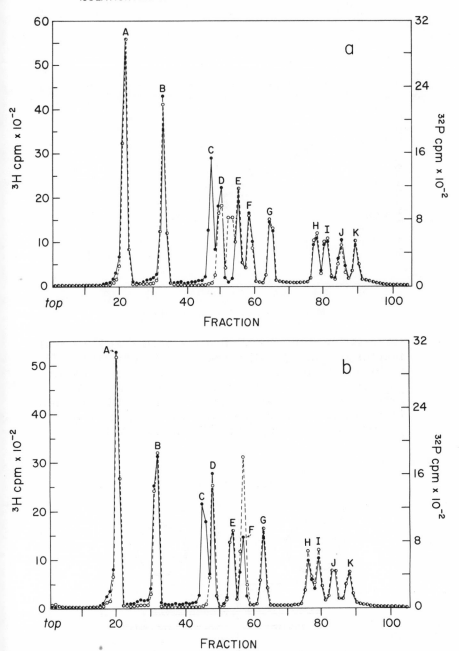

4. Polyacrylamide gel electrophoresis patterns of the Hind II + III fragments produced from a mixture of ^{32}P-WT DNA (•——•) and ^{3}H-mutant DNA (o---o). 4A is the pattern with mutant dl-801 and 4B from mutant dl-805.

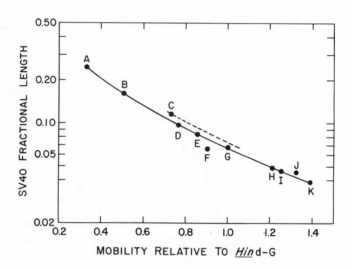

5.   Correlation of electrophoretic mobility and molecular
     lengths of Hind II + III fragments of SV40 DNA.   The
     mobility of each fragment is expressed relative to
     Hind fragment G.   The molecular lengths were
     estimated from the fraction of the total DNA present
     in each peak from at least ten different electro-
     phoretic analyses with the ^{32}P-labeled WT DNA.

MAPPING OF DELETION MUTANTS

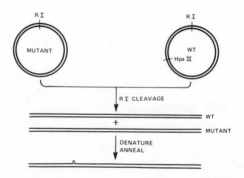

6.   Mapping of deletions of the HpaII restriction site of
     SV40 DNA.

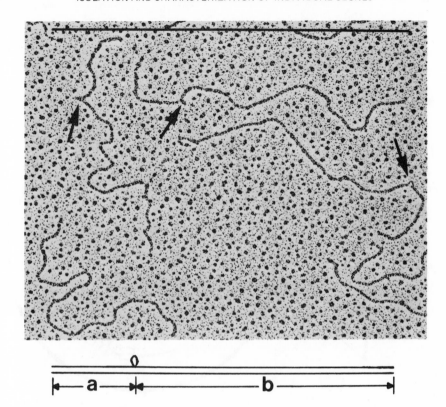

7.   Electron micrograph of heteroduplexes formed as
     illustrated in Fig. 6.   The arrows point to the puta-
     tive heteroduplexes containing the deletion loops.

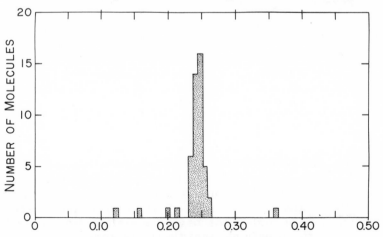

Figure 8. See legend on p. 12.

11

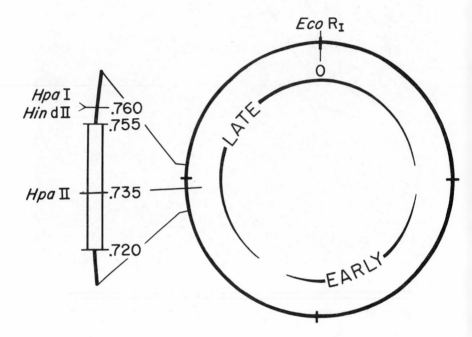

9. SV40 DNA map location of deletions which remove the HpaII restriction site and cause altered plaque morphology.

8. Location of the deletion loop in heteroduplexes between EcoRI endonuclease-cut DNA from mutant dl-810 and WT SV40 DNA. The histogram presents the number of molecules having a visible discontinuity along the length against their distance from the nearer end (segment a in Fig. 7).

Based on the location of the deletion loop (Fig. 8) and the estimate of the size of the deletion from the shortening of the Hind C fragment, the map coordinates of the deletions can be deduced (Fig. 9). Mutant dl(pm)-810 is lacking the segment of DNA from 0.720 to 0.755. This assignment is consistent with the fact that the HindII- and HpaI-restriction sites at 0.760 are still present in all the mutants.

At present we cannot comment on the physiologic defect caused by these small deletions. Because both deletions of existing DNA or insertions of new DNA (see next section) at this site still yield viable virions, we are inclined to believe that this region does not code for an essential protein. Further studies to clarify this point are in progress.

Construction of Insertion Mutants in Vitro

We have developed a general procedure for constructing mutants of SV40 virus by covalently inserting a short segment of extraneous DNA into the viral DNA molecule at specific locations. The supposition is that such an insert interrupts the normal sequence of nucleotides and either partially or completely inactivates the genetic function of that region. Moreover, the map position of such inserts should be readily discernible by heteroduplex analysis.

The overall protocol for constructing such molecules in vitro is a modification of the earlier procedure of Jackson, Symons and Berg (30)(see that paper for details). In this procedure, however, the homopolymer chains themselves serve as the insert (Fig. 10).

Briefly the procedure involves the following steps: a) opening of the SV40 circular DNA by a double-strand scission at a single location (in the example to be discussed here, by the HpaII restriction endonuclease); b) treatment with λ phage 5'-exonuclease to remove about 25 nucleotides from each 5' terminus; c) reaction of the product with either dATP or dTTP and the purified terminal transferase enzyme to attach short (about 50 residues) homopolymer "tails" of dA or dT to the 3' termini; d) denaturation and annealing of an equimolar mixture of dA- and dT-ended duplex molecules to generate molecules containing a dA-tail at one 3'-end and a dT-tail at the other 3'-end of the same duplex; e) cyclization of these linear molecules by their newly created cohesive ends and covalent closure of

13

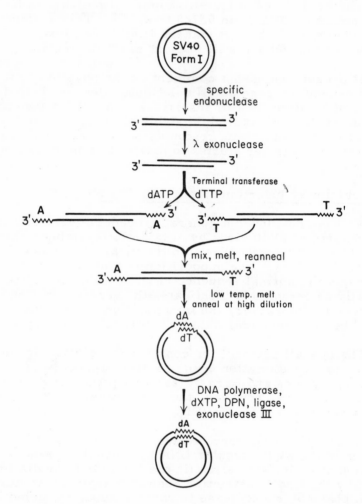

10.   Reaction scheme for the synthesis of covalently closed circles of SV40 DNA containing an insert of poly dA:dT.

the hydrogen-bonded circular molecules mediated by DNA polymerase, the four deoxynucleoside triphosphates and DNA ligase. By limiting the extent of addition of dA and dT "tails" to about 50 nucleotides, the final product is still small enough to permit encapsidation of the modified DNA molecule and therefore its propagation in virions.

Not unexpectedly, insertion of the short dA:dT sequence at the HpaII restriction site renders the circular molecule resistant to cleavage by HpaII endonuclease. Although our proof that there is an insert of dA:dT at 0.735 map position is not complete, heteroduplex analysis strongly supports this contention. The principle for mapping the poly dA:dT insertion is similar to that used to detect the small deletions at the HpaII restriction site (see Fig. 6). The molecules containing the dA:dT inserts can be cut to linear molecules with EcoRI endonuclease and heteroduplexed to EcoRI-cut WT DNA. Such heteroduplexes should contain a small loop of single-stranded dA and others a loop of dT (of course some of the reannealed products are homoduplexes of two WT or two modified strands). However, the small size of the insertion loops makes their direct visualization extremely unlikely. The dA loops could be visualized, however, by annealing linear SV40 molecules having dT tails (those shown in Fig. 10) to the heteroduplexes. This yields branched molecules (Fig. 11) in which the unit length homoduplex is attached by dA:dT base pairs to the dA insertion loop of the heteroduplex. Measurements of the three arms (see Fig. 11) of the branched molecule indicate that a unit length SV40 linear DNA molecule's poly dT tail can anneal to the heteroduplex at the position of the dA:dT insertion (0.735 map position).

Molecules containing the dA:dT insert are infectious and produce plaques in the absence of helper virus. The plaque morphology and rate of plaque development is virtually indistinguishable from those produced by the naturally occurring variants which lack this region.

With SV40 linear DNA molecules produced by random scissions (such as occurs with pancreatic DNase in the presence of Mn^{2+} (31)), and a modification of this synthetic procedure, we are optimistic that insertions or deletions can be introduced throughout the viral DNA molecule. Such synthetic defectives should be quite useful for mapping and functional identification of SV40 genes.

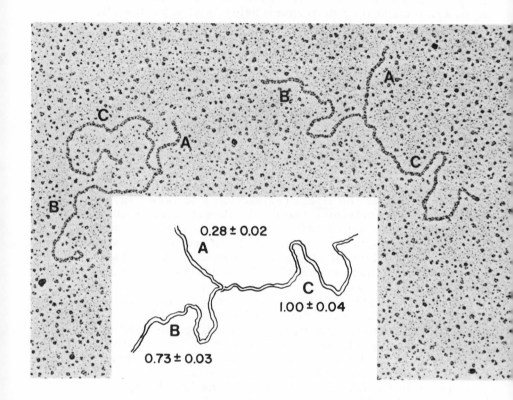

11. Electron microscopic visualization of small poly dA insertion loops present in heteroduplexes. Segment A and B comprise the linear heteroduplex formed from the EcoRI endonuclease-cut WT DNA and the EcoRI-cut DNA containing the poly dA:dT insert at the HpaII restriction site (0.735 map position). Segment C is the linear SV40 DNA molecule bound to the poly dA loop by its poly dT "tail".

16

ACKNOWLEDGEMENTS

This work was supported by research grants from the National Institutes of Health (GM 13235-09) and from the American Cancer Society (VC 23C).

J.E.M. is a trainee of the National Institutes of Health (5 TI GM 196-15).

J.C. is a member of the Department of Biological Sciences, University of California, Santa Barbara, California, and was supported in part by NSF 23429.

References

1. J.A. Robb and R.G. Martin, Virol. 41, 751 (1970).
2. P. Tegtmeyer and H.L. Ozer, J. Virol. 8, 516 (1971).
3. G. Kimura and R. Dulbecco, Virol. 49, 394 (1972).
4. S. Kit, S. Tokuno, K. Nakajima, D. Trkula and D.R. Dubbs, J. Virol. 6, 286 (1970).
5. J.Y. Chou and R.G. Martin, J. Virol., in press (1974).
6. J.A. Robb and R.G. Martin, J. Virol. 9, 956 (1972).
7. J.A. Robb, H.S. Smith and C.D. Scher, J. Virol. 9, 969 (1972).
8. P. Tegtmeyer, J. Virol. 10, 591 (1972).
9. J.F. Morrow and P. Berg, Proc. Nat. Acad. Sci. U.S.A. 69, 3365 (1972).
10. R. Davis, M. Simon and N. Davidson In Methods in Enzymology, ed. L. Grossman and K. Moldave (Academic Press, New York), Vol. 21, 413 (1971).
11. H.O. Smith and D. Nathans, J. Mol. Biol. 81, 419 (1973).
12. R.N. Yoshimori, Ph.D. dissertation, University of California, San Francisco Medical Center (1971).
13. C. Mulder and H. Delius, Proc. Nat. Acad. Sci. U.S.A. 69, 3215 (1972).
14. J. Hedgpeth, H.M. Goodman and H.W.P. Boyer, Proc. Nat. Acad. Sci. U.S.A. 69, 3448 (1972).
15. G.H. Sack, Jr. and D. Nathans, Virol. 51, 517 (1973).
16. P.A. Sharp, B. Sugden and J. Sambrook, Biochem. 12, 3055 (1973).
17. J.F. Morrow and P. Berg, J. Virol. 12, 1631 (1973).
18. K. Danna and D. Nathans, Proc. Nat. Acad. Sci. U.S.A. 68, 2913 (1971).
19. K.N. Subramanian, R. Dhar, J. Pan and S. Weissman, personal communication.

20. K. Yoshike, Virol. 34, 391 (1968).
21. K. Yoshike and A. Furuno, Fed. Proc. 28, 1899 (1969).
22. H. T. Tai, C. A. Smith, P. A. Sharp and J. Vinograd, J. Virol. 9, 317 (1972).
23. A. J. Girardi, Proc. Nat. Acad. Sci. U.S.A. 54, 445 (1965).
24. R. Radloff, W. Bauer and J. Vinograd, Proc. Nat. Acad. Sci. U.S.A. 57, 1514 (1967).
25. J. F. Morrow, P. Berg, T. Kelly and A. M. Lewis, Jr., J. Virol. 12, 653 (1973).
26. J. Sambrook, B. Sugden, W. Keller and P. A. Sharp, Proc. Nat. Acad. Sci. U.S.A. 70, 3711 (1973).
27. G. Khoury, M. A. Martin, T. N. H. Lee, K. J. Danna, and D. Nathans, J. Mol. Biol. 78, 377 (1973).
28. D. Nathans and K. J. Danna, Nature, New Biol. 236, 200 (1972).
29. G. C. Fareed, C. F. Garon and N. P. Salzman, J. Virol. 12, 484 (1972).
30. D. A. Jackson, R. H. Symons and P. Berg, Proc. Nat. Acad. Sci. U.S.A. 69, 2904 (1972).
31. E. Melgar and D. A. Goldthwait, J. Biol. Chem. 243, (1968).

ADENOVIRUS DNA:
TRANSCRIPTION DURING PRODUCTIVE INFECTION,
INTEGRATION IN TRANSFORMED CELLS, AND
REPLICATION IN VITRO.

MAURICE GREEN, TADASHI YAMASHITA, KARL BRACKMANN,
KEI FUJINAGA, MAX ARENS, WERNER BÜTTNER,
G. SHANMUGAM AND S. BHADURI

Institute for Molecular Virology, St. Louis University
School of Medicine, St. Louis, Missouri

Introduction. The DNA containing human adenoviruses
provide excellent models for studying the molecular
biology of the mammalian cell and the mechanisms that
regulate cell growth. Infection with adenoviruses may
result in either (i) productive infection infection in
which several hundred thousand virus particles are
formed and the cell is killed, or (ii) cell transformation
in which virus does not replicate but a portion of the
viral genome is integrated and cell growth is regulated
in an unknown manner by integrated viral genes. The
synthesis of adenovirus macromolecules during productive
infection mimics that of the host cell. Like cell DNA,
viral DNA is replicated in the cell nucleus, and is
transcribed to polycistronic RNA molecules that are
processed and transported to the cytoplasm for translation
in polyribosomes. The ability to identify viral DNA,
viral mRNA, and viral proteins facilitates the investigation of the mechanism of DNA replication and RNA
transcription and translation in mammalian cells using
viral macromolecules for analysis. Furthermore, late
after infection (18 hrs), cell DNA synthesis is blocked,
thus facilitating the analysis of viral DNA replication.
At this time, the synthesis of host cell mRNA, ribosomal
RNA, and host cell proteins are also inhibited, thus
facilitating the analysis of viral mRNA and protein
synthesis (5).
    In this paper, we present the results of recent
studies from our laboratory concerning (i) the regulation
of gene transcription in adenovirus 2 infected KB cells
by cellular and viral-imposed mechanisms; (ii) the
determination of the number and size of viral DNA
fragments integrated into adenovirus transformed cells;

and (iii) the replication of adenovirus DNA in vivo and
by a nuclear membrane complex in vitro.

Post-transcriptional processing of viral gene
transcripts by cellular mechanisms early during productive
infection of KB cells by adenovirus 2. Studies on the
molecular events in adenovirus replication often use as
a model adenovirus 2 infected suspension cultures of
human KB cells (6). In this system, mature virus is
formed from 13 to 24 hrs after infection (Fig. 1) and
viral DNA synthesis begins at 6 to 7 hours. Early
viral gene functions and early viral mRNA synthesis
occur prior to this time, while late viral gene functions
and late mRNA synthesis occur after 6-7 hrs. Hybridization experiments with viral DNA immobilized on nitrocellulose filters have revealed that 80-100% of the
adenovirus 2 genome is transcribed by 18 hrs after
infection (4), and that 8-20% of viral RNA sequences
synthesized present late after infection are also found
early after infection (3). These experiments performed
by conventional DNA-RNA hybridization measurements,
detect stable (long-lived) viral RNA species, and most
likely would miss viral RNA species that turned over
rapidly, and thus were present in low concentrations.

In order to detect possible short-lived precursor
viral RNA molecules, and to measure quantitatively the
fraction of the viral genome represented by viral RNA
species present in different relative abundance, we
performed highly sensitive RNA-driven hybridization
reactions using highly radioactive adenovirus L and H
DNA strands as hybridizing probe. To prepare L- and H-
DNA strands, denatured viral DNA was complexed with
poly(U,G) and centrifuged to equilibrium in CsCl gradients
(Fig. 2). Separated peaks of H- and L-strand complexes
were rebanded (Fig. 2) and purified further by sedimentation in alkaline sucrose gradients. Purified labeled
L- and H-DNA strands were annealed with excess RNA
isolated from adenovirus infected cells early after
infection (5 hrs) and the fraction of DNA that became
hybrid after various times was measured with the
single-stranded nuclease, S_1. The data are plotted in
Figure 3 as RNA C_0t (moles of RNA nucleotide/liter
multiplied by time in seconds) against percent DNA
hybridized (on a log scale). Since first order kinetics
are expected for the hybridiztion of excess RNA with
tracer amounts of DNA, each class of viral RNA should
yield a straight line with a slope proportional to its

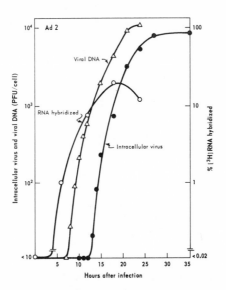

Figure 1. Time course of formation of intracellular virus, viral DNA, and viral mRNA.

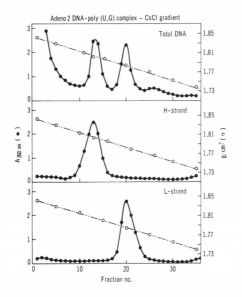

Figure 2. Resolution of adeno 2 L and H strand DNA as complexes with poly(U,G) by equilibrium centrifugation in CsCl gradients. Upper panel - centrifugation of denatured adenovirus 2 DNA after annealing with poly(U,G). Middle panel - rebanding of H strand. Lower panel - rebanding of L strand.

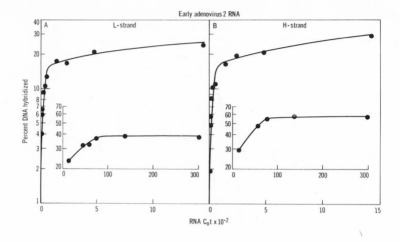

Figure 3. Hybridization of ^{32}P-labeled H and L DNA strands with early (5 hr) adeno 2 infected cell RNA.

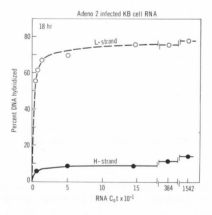

Figure 4. Hybridization of ^{32}P-labeled H and L-strand DNA with late (18 hr) adeno 2 infected cell RNA.

relative abundance. From Figure 3, two RNA classes are evident: (i) abundant RNA which hybridizes rapidly and saturates about 14% of labeled DNA at RNA C_0t values of about 100, and (ii) scarce RNA which hybridizes very slowly and reaches saturation only at RNA C_0t values of 10,000 or higher. At saturation, about 55% of the H-strand and 40% of the L-strand have formed a hybrid with RNA. Thus the equivalent of the entire viral genome or perhaps slightly less (95%) are transcribed early after infection. The two classes of viral RNA present at 5 hours after infection probably represent stable viral RNA (abundant RNA) and short-lived primary transcripts (scarce RNA) that are precursors to abundant RNA. The two classes are presently at a molar ratio of 180 to 1, as determined from a computer analysis of the data (2). Similar classes of RNA were found in cells infected in the presence of cycloheximide, an inhibitor of protein synthesis, indicating that this transcriptional pattern represents a cellular regulatory mechanism (Green, Brackmann, Cartas, and Devine, unpublished data). We conclude that a major mechanism of gene regulation in human cells is the post-transcriptional processing of complete transcripts of DNA molecules.

Viral imposed regulation of gene expression late after infection. A dramatic change occurs in transcription late after infection. Only abundant viral RNA transcripts are detected, and these are derived from about 80% of the L strand and only 10% of the H-strand (Fig. 4). Similar patterns were found from 9 to 36 hrs after infection (Green, Brackmann, Cartas, and Devine, unpublished data). Several questions are raised by these observations: (i) what prevents the transcription of a major segment of the H-strand late after infection? (ii) how is the L strand RNA transcript protected from degradation by cellular nucleases whose function presumably are the destruction of the same RNA transcripts early after infection? (iii) how is the switch from early to late viral gene transcription achieved? The answer to these questions will enhance our understanding of how mammalian cells regulate gene expression and how viral genes alter cellular control.

Isolation of DNA-strand specific early mRNA species from human adenovirus 2 infected cells. The adenovirus genome codes for only a small number of early proteins, among which could be viral protein(s) responsible for cell transformation. It is therefore

of interest to characterize the early viral mRNA species
that code for these proteins. To achieve this end, we
annealed RNA from polyribosomes labeled from 4 to 7 hrs
after infection in the presence of cycloheximide
(cycloheximide increases the concentration of viral
mRNA (9)) with L and H DNA strands under mild conditions.
As shown in Fig. 5, three size classes were detected by
hybridization with unfractionated viral DNA, a major
19-20S RNA and minor 21-26S and 15-18S broad RNA peaks,
as previously observed by Parsons and Green (9).
However, selection with individual DNA strands showed
that viral 19-20S RNA consists of two relatively
homogeneous RNA species with slightly different mobilities,
L and H strand-specific mRNA with molecular weight of
0.74×10^6 and 0.77×10^6 respectively (Fig. 5). The
broad 15-18S RNA peak is derived from the H-strand and
the broad 21-26S RNA peak from the L strand. The
nature of these two minor components is unknown. The
two major viral mRNA molecules can now be used in
further studies to map their origin on the viral genome
and for translation in cell-free protein synthesizing
systems to establish their protein gene products.

Integration of viral genome fragments in cells
tranformed by adenoviruses. Cells transformed by
human adenovirus contain neither infectious virus nor
the entire viral genome. The strongest evidence that
viral DNA is integrated is the isolation of polycistronic
RNA molecules containing viral and cell sequences (11).
The number of partial viral DNA genomes present in
cells transformed by DNA viruses is difficult to establish
by hybridization methods previously described. Hybridization of transformed cell DNA with radioactive viral
cRNA (synthesized by copying adenovirus DNA with the E.
coli RNA polymerase) has estimated 90-100 viral DNA
equivalents in adeno 7 transformed cells and 40-50 DNA
equivalents in adeno 12 transformed cells (7). But,
the unknown viral sequence content of viral cRNA and
transformed cell DNA limits the accuracy of such
measurements.

The standard plot of DNA reassociation kinetics
described by Britten and Kohne (1), often used to
measure viral and cell DNA copies in eukaryotic cells,
leads to serious errors when used to estimate copies of
partial viral genomes in transformed cell DNA. For
example, similar 50% reassociation values ($C_{o}t_{1/2}$)
would be obtained for 200 copies of 20% of the adeno-

24

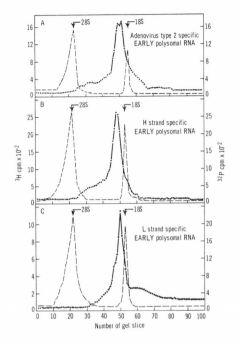

<u>Figure 5.</u> Polyacrylamide gel electrophoresis of DNA
strand-specific early viral mRNA molecules. Early
polyribosomal RNA labeled with ^{3}H-uridine was annealed
to unfractionated adenovirus 2 DNA (a), H-strand DNA
(b), or L-strand DNA (c). Viral RNA was eluted and
electrophoresed on agarose-polyacrylamide gels. [32]P-
labeled RNA served as molecular markers •———•: ^{3}H-cpm
in virus-specific RNA, ------: ^{32}P-cpm in ribosomal
marker RNA (From Büttner, Veres-Molnar, and Green, in
manuscript).

virus genome as for 5 copies of the entire viral genome
(Fujinaga, Sekikawa, Yamazaki, and Green, in manuscript).

To overcome the above difficulties, Fujinaga,
Sekikawa, Yamazaki, and Green (in manuscript) have
developed a new mathematical treatment of viral DNA
reassociation kinetic data that estimates both the
fraction of the viral genome and the copy number from
the initial rate of reassociation of labeled viral DNA
in the presence of transformed cell DNA. The theoretical
and experimental basis of this analysis was validated
by reconstruction experiments in which the reassociation
of labeled adeno 7 DNA was analyzed in the presence of
unlabeled partially homologous adeno 2 DNA (Fujinaga,
Sekikawa, and Yamazaki, in manuscript).

The reassociation of adeno 7 DNA in the presence
of adeno 7 transformed cell DNA (5728 clone 15 and
uncloned 5728 cell lines) is shown in Figure 6. The
data are plotted as C_{ss}/C_{ds} against the reciprocal of
C_0t (C_{ss} and C_{ds} are the concentration of single and
double-stranded DNA, respectively, and C_0t is the
concentration of DNA in mole/liters x time in seconds).
This modification of the standard plot of Britten and
Kohne (1) is convenient for the analysis of single
molecular species, providing a sensitive indicator of
second order kinetics, and permitting the analysis of
copy number and fraction of related viral gene sequences.
As shown in Figure 6, the straight line plot for the
reassociation of adeno 7 DNA in the presence of E. coli
DNA clearly indicates that second order reassociation
kinetics are followed, as expected when viral DNA
sequences are present in equimolar amounts, i.e.
nonreiterated. In contrast, when labeled adenovirus 7
DNA is reassociated in the presence of transformed cell
DNA, a marked deviation from linearity is obtained at
low values of $1/C_0t$ (i.e. large C_0t values). From the
slope of the initial rate of the reaction, the number
of copies of viral DNA segments and the fraction of the
viral genome are calculated as described by Fujinaga,
Sekikawa, Yamazaki, and Green (in manuscript). From
these data and several repeat experiments, we estimate
that there are 280 to 310 copies per cell of 18 to 20%
of the viral genome in adeno 7 transformed cells.
Similar measurements with adeno 12 transformed cells
revealed about 40-60 copies of 40-60% of the viral
genome. These data were confirmed by independent
measurements of the reassociation kinetic of adeno 7

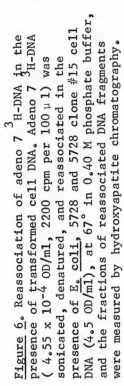

Figure 7. Association of newly synthesized DNA with the nuclear M-band. Nuclei prepared from 18 hr infected KB cells pulse labeled with ^{3}H-thymidine for 5 min were treated with Na dodecyl sarcosinate in the presence of $MgCl_2$, and centrifugated at 20,000 rpm for 20 min in a Spinco SW-41 rotor onto a two layer discontinuous gradient consisting of 15% and 40% sucrose.

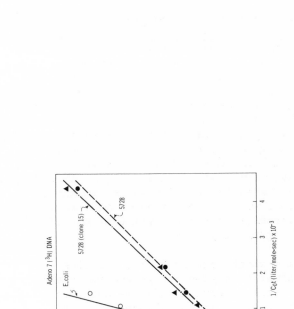

Figure 6. Reassociation of adeno 7 ^{3}H-DNA in the presence of transformed cell DNA. Adeno 7 ^{3}H-DNA (4.55 x 10^{-4} OD/ml, 2200 cpm per 100 μ l) was sonicated, denatured, and reassociated in the presence of E. coli, 5728 and 5728 clone #15 cell DNA (4.5 OD/ml), at 67° in 0.40 M phosphate buffer, and the fractions of reassociated DNA fragments were measured by hydroxyapatite chromatography.

DNA fragments that were obtained by treatment of adeno 7 DNA with Hemophilis influenza endonuclease Rd (Sekikawa, Odajima, and Fujinaga, unpublished data).

Our data indicate that there are large numbers of specific viral DNA fragments present in hamster cells transformed by adeno 7 and adeno 12. Sharp, Phillipson, and Sambrook (personal communication) have found approximately two copies of about 40-50% of the viral genome in adeno 2 transformed cells. The basic mechanism of integration may differ in cells transformed by these human adenoviruses of different oncogenic groups, or may be influenced by the conditions of transformation. It is interesting that adeno 7 and adeno 12 belong to weakly and highly oncogenic group B and A respectively, while adeno 2, a member of group C, transformed rat embryo cells in vitro but has not induced tumors in nonconditioned newborn hamsters.

DNA replication complex from adenovirus 2 infected KB cells and associated DNA binding proteins. Cells productively infected with human adenoviruses are excellent models for analyzing the replication of a linear duplex DNA molecule in a mammalian cell. To localize a site of viral DNA synthesis, adenovirus 2 infected KB cells were pulse-labeled for 5 minutes with ^{3}H-thymidine at 18 hrs after infection and nuclear membrane fractions were prepared by both the M-band and discontinuous sucrose gradient methods. The M-band procedure (10) isolates a membrane fraction which contains most of newly synthesized viral DNA, as shown in Figure 7. Similar association of newly synthesized viral DNA was observed with a nuclear membrane fraction isolated by the discontinuous sucrose gradient method (8) at density 1.18-1.20 g/ml (13).

Based on studies which indicated that proteins synthesized early after infection are needed for viral DNA synthesis late after infection (13), we analyzed the polypeptides synthesized in the absence of viral DNA synthesis that became associated with nuclear membrane fractions isolated from cells late after infection. M-band preparations from virus-infected cells labeled with ^{3}H-leucine were coelectrophoresed with similar preparations from ^{14}C-leucine labeled uninfected cells. Two major peaks with molecular weights of about 75,000 and 45,000 and lesser quantities of two smaller polypeptides were found in a nuclear membrane fraction from infected cells (Fig. 8); these

28

were not detected in the appreciable quantities in
uninfected cells (Fig. 8). These polypeptides are
similar to the DNA-binding proteins isolated from
adeno 5 infected monkey kidney cells (12). We therefore
studied the DNA-binding properties of the proteins
components associated with the nuclear membrane fraction
from adeno 2 infected cells as isolated by the discontinuous sucrose gradient method. The components of the
nuclear membrane fraction isolated from infected and
uninfected cells were fractionated on single-stranded
calf thymus DNA cellulose columns by stepwise NaCl
elution and each eluate was analyzed by electrophoresis
on 6% polyacrylamide gels. From 50-60% of proteins
present in infected cell membrane and 40-50% of those
present in the uninfected cell membrane bound to DNA
cellulose (Shanmugan, Bhaduri, Arens, and Green, in
manuscript). As shown in Figure 9, electrophoresis of
the 0.6 M eluate showed the same proteins of molecular
weight 75,000 and 45,000 that were detected in the DNA
replication complex. These proteins were absent from
the DNA binding proteins from the nuclear membrane and
from the cytoplasm of uninfected cells (Shanmugan,
Bhaduri, Arens, and Green, in manuscript). The functions
of these two DNA binding proteins is not known. The
nuclear membrane fraction contains, in addition, DNA
polymerase and endonuclease activity, and will synthesize
adenovirus DNA sequences in vitro. We operationally
refer to this fraction as "DNA replication complex",
and below we discuss some of its properties.

The DNA replication complex from 18 hr infected
cells synthesizes adenovirus DNA sequences in vitro.
As shown in Table I, in vitro synthesized DNA hybridized
with viral DNA with the same efficiency as did authentic
radioactive adenovirus 2 DNA. No significant hybridization was detected between in vitro DNA and KB cell DNA
(Table I). Thus the viral specificity of DNA synthesis
in vivo is maintained in vitro.

Maximal DNA polymerase activity occurred in vitro
at pH 8. KCl and ATP are not essential but
stimulate polymerase activity. The optimal Mg^{2+}
concentration is 7 to 20 mM, and Mn^{2+} is a poor substitute.
All four deoxyribonucleoside triphosphates were incorporated, and DNA synthesis was inhibited by sodium
pyrophosphate, DNase, N-ethylmaleimide, and actinomycin
D. RNase inhibited about 50%, suggesting a possible
involvement of RNA in DNA synthesis. DNA was synthesized

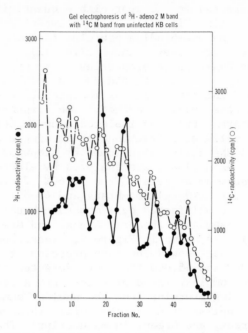

Gel electrophoresis of ^{3}H - adeno 2 M band
with ^{14}C M band from uninfected KB cells

Figure 8. Coelectrophoresis on Na dodecyl SO$_4$ gels of ^{3}H-leucine labeled M-band polypeptides from adenovirus 2 infected KB cells with ^{14}C-leucine labeled M-band polypeptides from uninfected KB cells.

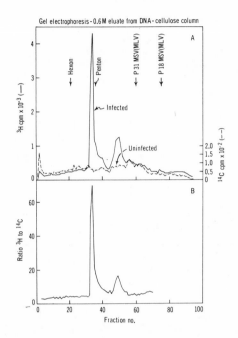

Figure 9. SDS-polyacrylamide gel electrophoresis of
DNA-binding proteins of replication complex mixture
eluted with 0.6 M NaCl. Fractions 74 to 82 of Figure
9B were pooled, precipitated with TCA and the precipitate
(144,300 ^{3}H cpm and 18,400 ^{14}C cpm) was analyzed on 6%
polyacrylamide gels. In Figure 9a, (_____), ^{3}H cpm and
(-----), ^{14}C cpm represent radioactivity of proteins of
infected and mock infected cells respectively. Figure
9b represents the data of Figure 9a plotted as the
ratio of ^{3}H (infected) to ^{14}C (mock infected) cpm.

TABLE I   Viral nature of in vitro synthesized DNA

Input DNA	Immobilized DNA (µg/filter)	Bound DNA cpm	%
Adenovirus 2 ³H-DNA (14,100 cpm)	adenovirus DNA (10 µg)	7,050	50
" "	KB cell DNA (10 µg)	0	0
In vitro ³H-DNA (4,350 cpm)	adenovirus DNA (10 µg)	2,090	48
" "	KB cell DNA (10 µg)	45	1
KB cell ³H-DNA	KB cell DNA (10 µg)	2,260	34

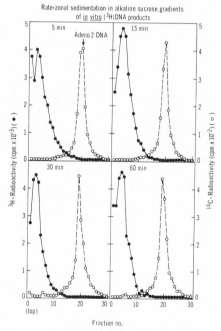

Figure 10. Rate-zonal sedimentation in alkaline sucrose gradient of ³H-DNA synthesized in vitro with the adenovirus 2 DNA replication complex. The standard assay was used with the incubation times noted. The size of DNA was analyzed as marked by alkaline sucrose gradient centrifugation with adenovirus 2 ¹⁴C-DNA as marker.

in vitro by a semiconservative mechanism (Yamashita, Arens, and Green, in manuscript).

The product synthesized from periods of time from 5 to 60 minutes sediments at about 6-7S in alkaline sucrose gradient (Fig. 10). Zone sedimentation in neutral sucrose gradients suggest that these DNA fragments are hydrogen bonded to 18S DNA. Since the replication complex contains an endonuclease which cleaves parental viral DNA to 18S DNA fragments (Yamashita, Arens, and Green in manuscript), it is not clear whether 18S DNA is a cleavage product or an intermediate in viral DNA synthesis. It is possible that the "DNA replication complex" represents only part of the machinery required for adenovirus DNA synthesis in vivo since complete adenovirus DNA molecules are not synthesized in vitro under our reaction conditions. Alternatively, it is possible that cofactors needed for DNA ligase activity are absent from the reaction mixture. Further analysis of the nuclear membrane fraction should further our understanding of the mechanism of DNA replication.

Summary. Early after productive infection of human KB cells by adenovirus 2, the transcription of viral genes is regulated by cellular mechanisms. Two populations of viral RNA molecules were detected by hybridization of highly radioactive viral L and H DNA strands with total cellular RNA in excess: (i) scarce RNA, representing the equivalent of the entire viral genome, (ii) abundant RNA, representing about 14% of each viral DNA strand, probably the end product of processing of scarce RNA by cellular mechanisms. Two major viral mRNA species were isolated from polyribosomes early after infection, one derived from the L strand with a molecular weight of 0.74×10^6 and the second from the H strand with a molecular weight of 0.76×10^6.

By analysis of the early kinetics of reassociation of denatured adenovirus DNA in the presence of unlabeled transformed cell DNA by a new mathematical treatment, we found evidence that approximately 300 copies of 18-20% of the adeno 7 genome and 50 copies of about 50% of the adeno 12 genome are present in adeno 7 and 12 transformed hamster cells, respectively.

Nuclear membrane fractions were isolated from adeno 2 infected KB cells late after infection and shown to contain DNA binding proteins of molecular weight 75,000 and 45,000. The nuclear membrane fraction isolated by the discontinuous sucrose gradient method can synthesize adenovirus DNA sequences in vitro. Some properties of this "DNA replication complex" and the in vitro DNA products are described.

REFERENCES

1. R. J. Britten, and D. E. Kohne, Science 161, 529-540, 1968.

2. N. Frenkel and B. Roizman, Proc. Natl. Acad. Sci. U.S. 69, 2654-2658, 1972.

3. K. Fujinaga, and M. Green, Proc. Natl. Acad. Sci. U.S., 65, 375-382, 1969.

4. K. Fujinaga, S. Mak, and M. Green, Proc. Natl. Acad. Sci. U.S., 60, 959-966, 1968.

5. M. Green, Ann. Rev. Biochem 39, 701-756, 1970.

6. M. Green, and G. E. Daesch, Virology 13, 169-176, 1961.

7. M. Green, J. T. Parsons, M. Pina, K. Fujinaga, H. Caffier, and M. Landgraf-Leurs, Cold Spring Harbor Sym. Quant. Biol. XXXV, 803-818, 1970.

8. D. M. Kashing, and C. B. Kasper, J. Biol. Chem. 244, 3786-3792, 1969.

9. J. T. Parsons, and M. Green, Virology, 45 154-164, 1971.

10. G. Y. Tremblay, M. J. Daniels, and M. Schaecter, J. Mol. Biol. 40, 65-76, 1969.

11. D. Tsuei, K. Fujinaga, and M. Green, Proc. Nat. Acad. Sci. 69, 427-430, 1972.

12. P. C. van der Vliet and A. J. Levine, Nature New Biol. 246, 170-173, 1973.

13. T. Yamashita, and M. Green, J. Virology, in press, 1974.

TRANSFORMATION BY HERPES SIMPLEX

AND CYTOMEGALOVIRUSES

Fred Rapp and Stephen C. St. Jeor

Department of Microbiology
The Milton S. Hershey Medical Center
The Pennsylvania State University
Hershey, Pennsylvania 17033

Cell transformation by a virus is associated with the permanent transfer of virus genetic material to a cell. A number of animal viruses, including certain members of the herpesvirus group, are able to transform mammalian cells (1-3). Transformation is often accompanied by a change in the cells from a normal to a malignant phenotype. Although a malignant change in a cell might represent loss rather than gain of genetic material, certain characteristics of malignant transformation indicate a permanent addition of genetic material. These include the acquisition of new antigens, the appearance of new or altered enzymes and/or the presence of virus specific nucleic acids.

Herpesviruses are known to be the etiologic agent for a variety of malignant diseases in lower animals. Marek's disease, a lymphoproliferative disorder in chickens, is almost certainly caused by a herpesvirus (4,5). An attenuated vaccine (6) has greatly decreased the incidence of this malignant disease indicating the association of the virus with malignant changes in its host. A renal adenocarcinoma (Lucké tumor) occurring in certain frogs (Rana pipiens) has long been thought to have a virus etiology (7). A herpesvirus has been observed under the electron microscope in both tumor biopsies as well as in cells derived from the tumor (8,9). The evidence implicating a herpesvirus in the Lucké tumor is still unclear because other viruses, including another herpesvirus and a papova-like virus, have been isolated from the tumor cells (10,11).

Herpesvirus saimiri and ateles produce lymphomas in lower primates (12,13). Other herpesviruses with oncogenic potential have been isolated from cottontail rabbits (14), guinea pigs (15) and possibly cattle (16).

The evidence in man for the involvement of herpesviruses

in neoplastic disease, although not as strong as for some of the lower animal systems, is becoming increasingly convincing. Epstein-Barr virus has been associated with Burkitt's lymphoma and less strongly associated with nasopharyngeal carcinoma (17,18). Other herpesvirus which might be potential oncogenic agents include herpes simplex virus types 1 and 2 and human cytomegalovirus.

Association of Herpes Simplex Viruses to Neoplasia in Man:

Recent laboratory and epidemiological observations have indicated that herpes simplex virus (HSV) may be involved in neoplasia in humans. Naib et al. (19) reported an association between the appearance of cervical anaplasia and the occurrence of genital herpes lesions. This observation was followed by epidemiological findings that patients with cervical cancer have a higher incidence of antibody against HSV-2 than matched control groups (20,21). More direct evidence for the involvement of HSV with cervical carcinoma comes from the observations that HSV-2 antigens are detected in exfoliated squamous carcinoma cells and that virus can be isolated from degenerating tumor cells (22). These findings were followed by the observation that a small amount of the DNA isolated from cervical cancer cells would hybridize with HSV-2 DNA (23). Recently, it was shown that patients with certain cancers contain complement fixing antibodies to herpesvirus non-virion (actually, labile) antigens whereas control patients are negative for complement fixing antibody against the same antigens (24,25). Although the antigens are claimed to be "non-virion" their origin and localization is unknown.

In Vitro Transformation with Herpes Simplex Viruses:

The ability of both HSV-1 and 2 to transform cells in vitro has been demonstrated in several laboratories. Perhaps the most convincing studies were carried out using the permanent transfer of genetic information to code for a virus enzyme or for virus antigens. It is known that both HSV-1 and 2 are capable of inducing thymidine kinase in their host cell. Furthermore, the induced enzyme appears to be virus specific in its properties (26,27). The enzyme induced by type 1 virus can be distinguished from the type 2 enzyme because of greater heat stability of the former. Davis et al. (28) demonstrated that mouse L cells deficient in the enzyme thymidine kinase can be converted to a positive thymidine kinase phenotype using HSV-1 or 2 which had been inactivated with ultraviolet light; the new thymidine kinase in the cells has the properties of the herpesvirus used to infect the cells. Furthermore, the DNA which codes for the enzyme is a genetically stable marker and remains with the progeny of the infected cells.

We have been investigating the possibility that herpes simplex virus is capable of in vitro malignant transformation of cells. Prior to these studies, attempts had been made to produce tumors in Syrian hamsters by the injection of herpes simplex virus (29). However, this procedure rarely yields tumors and herpesvirus antigens have not been detected in the transformed cells. The L cells used by Munyon and his colleagues also have the disadvantage that they are malignant prior to virus exposure.

Initially, we attempted to transform hamster embryo fibro-blasts (HEF) cells using HSV-2. A problem in these studies was the fact that HEF cells are permissive for HSV and the cells were destroyed before transformation could be observed. It had been reported that the ability of a virus to form a plaque following exposure to UV light is more rapidly lost than is the ability to transform a cell (30,31). Plaque formation is dependent upon the expression of a large number of virus genes whereas the transformation of a cell is the result of the expression of a relatively small amount of genetic information. Considering these reports in the initial studies, UV inactivated HSV-2 was used as a trans-forming virus. Confluent monolayers of HEF cells were infected with UV inactivated HSV-2. Following virus adsorp-tion, the cells were trypsinized and placed into separate flasks. Two of 17 of original cultures developed foci of morphologically transformed cells within 21-28 days after infection. The cells were grown into a cell line and inject-ed into inbred LSH hamsters. Tumors were detectable 10-16 weeks after inoculation. Both the original cells and the tumor cells contained antigens specific for HSV-2. In further studies, we were able to demonstrate that the sera from hamsters with tumors reacted with HSV-1 and 2 infected cells and were capable of neutralizing HSV-1 and 2. Further-more, the sera from the tumor bearing hamsters reacted with the original transformed cells (32). Subsequent to these studies, Collard et al. (33) isolated HSV-2 specific mRNA from the transformed cells.

As an extension of the cell transformation studies by UV inactivated viruses, HSV-1 was examined for its oncogenic

37

potential in HEF cells. The procedure was basically identical to the method used to transform cells using HSV-2.

Clinical isolates of HSV-1 were tested for transforming potential in HEF. Two of twelve strains tested induced the development of five or more transformed foci in at least 1 of 3 tests. The morphology of the transformed cells induced by HSV-1 differed from the earlier observations with the HSV-2 transformed cells. The HSV-2 cells had been predominantly fibroblastic cells; however, the HSV-1 transformed cells were of epithelial morphology (34). Figure 1 is a photomicrograph of the HSV-1 transformed cells and depicts the epithelial morphology present in these cells. HSV-1 transformed cells reacted with anti-HSV-1 sera using a fluorescent antibody test, indicating the presence of HSV-1 specific antigens in the transformed cells.

The cells transformed by HSV-1 were tested for oncogenic transformation in newborn hamsters. One of two cell lines transformed by HSV-1 produced tumors from 3-11 weeks after injection. The tumors were not of the fibrosarcoma type as seen in the HSV-2 transformed cells but resembled an adenocarcinoma. This is particularly important as the majority of solid tumors in humans are carcinomas rather than fibrosarcomas. This system is the first example of a DNA virus capable of transforming cells which can produce a carcinoma in laboratory animals.

Although these studies indicated that HSV-1 and 2 have the potential to transform mammalian cells if their lytic function is lost, the frequency with which a human population would be exposed to HSV which lacks this function is unknown. It has been shown by a number of investigators that in the presence of light, heterocyclic dyes, such as neutral red, are capable of inactivating viruses (35). As these dyes are used currently for the treatment of localized herpes simplex infections (36), it was important to determine if cell transformation and/or oncogenic transformation could be accomplished using neutral red inactivated herpes simplex virus (37).

These studies were carried out with the following DNA containing animal viruses: HSV-1, HSV-2 and SV40 virus. SV40 was included because it is a well characterized mammalian tumor virus and thus was an excellent positive control.

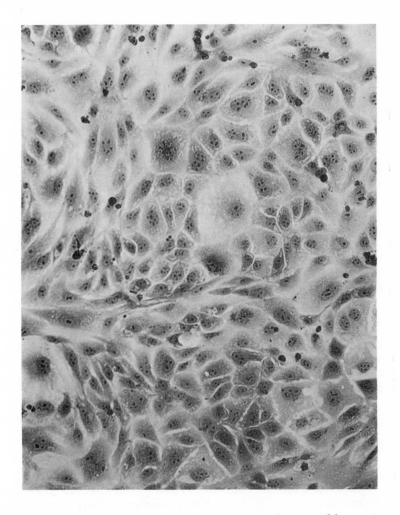

Fig. 1. Photomicrograph of hamster embryo cells transformed with HSV-1. The cells are stained with hematoxylin
and eosin and have a predominantly epithelial morphology.

Figure 2 is a photomicrograph illustrating the results of an indirect immunofluorescence test of HEF cells transformed with neutral red inactivated HSV-1. The positive serum was obtained from hamsters bearing an HSV-2 tumor. The data presented in these studies, as presented in Table 1, indicate that DNA-containing viruses are capable of transforming cells after photodynamic inactivation using neutral red.

The transformed cells were tested for oncogenicity in baby hamsters. Both the HSV-1 and 2 cells were capable of inducing tumors. To date, the HSV-1 tumors have not been tested but the HSV-2 tumors contain specific virus antigens. Sera from the tumor bearing animals react with both the transformed cells as well as with cells infected with HSV-2.

Work from a number of laboratories has confirmed our initial reported findings relating to the oncogenic capabilities of herpes simplex virus. Kutinová et al. (38) used inactivated HSV to convert a weakly malignant established cell line to a highly malignant state. Darai and Munk (39) demonstrated that if human embryonic lung cells infected with HSV were maintained at a non-permissive temperature, cell transformation occurred. Personal communications from a number of other investigators (Schaffer, McNaab, McAuslan, Wildy, Casto) have further clarified the capability of these viruses to transform mammalian cells in vitro.

Cell Transformation By Cytomegalovirus:

It has been known for some time that human CMV is often isolated from patients with neoplastic diseases (40,41). It is also well recognized that CMV can produce latent infection in its host (42,43).

In vitro evidence of cell transformation has been reported by Lang et al. (44). They found that CMV is capable of inducing an abortive transformation of human fibroblasts in soft agar. Because of these findings and considering the oncogenic potential of HSV-1 and 2, we attempted in vitro transformation studies with human CMV. The method used for transformation of cells by human cytomegalovirus was essentially the method used for the herpes simplex virus studies. It was known that in vitro replication of human CMV is limited to fibroblastic cells from the animal species from which the virus was isolated; however, there was ample evidence from observations in our own laboratory that CMV

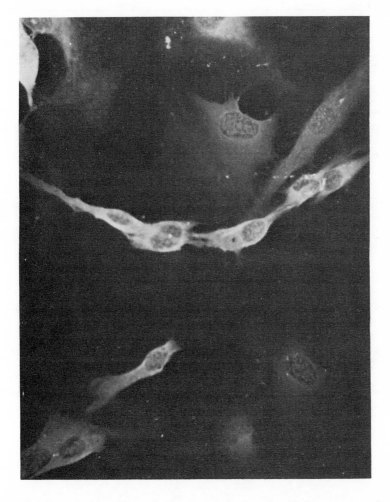

Fig. 2. Immunofluorescence micrograph of hamster embryo
cells transformed by neutral red inactivated HSV-1. Antiserum used in this test was obtained from hamsters with
HSV-2 tumors.

Table 1. Detection of Virus Antigens by Fluorescent Antibody in Cells Transformed by Neutral Red Inactivated Virus.

Cell Type	Anti-Sera			
	HSV-1[a]	HSV-2[a]	HSV-2[b] Tumor	SV40[c]
HSV-1[d]	+	+	+	−
HSV-2[d]	+	+	+	−
SV40[d]	−	−	−	+
Normal hamster embryo fibroblast	−	−	−	−
HEF + HSV-1[e]	+	+	+	−
HEF + HSV-2[e]	+	+	+	−
333-8-9[f]	+	+	+	−

[a]Pooled sera from hamsters immunized against HSV-1 or HSV-2.

[b]Pooled sera from hamsters bearing tumors induced by UV inactivated HSV-2 cells.

[c]Pooled sera from hamsters with SV40 induced tumors.

[d]Hamster embryo fibroblasts transformed by either neutral red inactivated HSV-1, HSV-2 or SV40.

[e]Hamster embryo fibroblasts infected with either HSV-1 or HSV-2.

[f]Hamster embryo fibroblasts transformed by UV inactivated HSV-2.

abortively infects nearly all cell types. There are three
markers we had observed indicative of an abortive infection
by CMV. These are the stimulation of cell DNA replication
(45), cytopathology (46,47), and induction of antigens
demonstrated by using a fluorescent antibody assay (47).
The block in CMV replication in these various cell types
appears to be at the level of DNA replication.

The same experimental design was used in these studies as
in our earlier investigations. Virus suspensions were
partially inactivated with UV light before addition to the
cell monolayers. Control cells were mock infected simultan-
eously with growth medium. Following adsorption of the virus,
the cells were observed for the development of foci. After
20 days incubation, 16 foci of non-contact inhibited cells
were observed. A single clone survived and after several
passages, a fibroblastic cell line was derived. The cells
were injected into both newborn and weanling Golden Syrian
hamsters and after 10 weeks, a tumor was observed.

Table 2 illustrates the results obtained in a study
designed to detect CMV antigens in various cell preparations
(48). The CMV transformed cells reacted with anti-CMV
convalescent sera when tested for both fixed and membrane
antigens. The cells derived from the original tumor failed
to react when tested for CMV internal antigens. However,
when they were examined for the presence of CMV surface antigens, positive results were obtained. In these studies, 47%
of the CMV transformed cells and 17% of the hamster tumor
cells were positive for surface antigens. Hamster embryo
fibroblasts infected with HSV-2 weakly reacted in a test for
surface antigens, indicating either shared antigens or that
the antisera used also contained HSV-2 antibodies.

It was of interest to determine whether the sera from
either the original tumor bearing hamster, or those animals
developing tumors following the injection of the tumor cells,
would react with cells infected with CMV. The sera from
animals bearing tumors reacted with cells infected with
either the C-87 or AD-169 strain of CMV. This same sera
failed to react with uninfected HEF and HEL cells or with
HEL cells infected with HSV-2.

Stimulation of Cell DNA By Human CMV:

Stimulation of cellular DNA is a characteristic common
to many oncogenic DNA viruses (49-53); however, its function

43

Table 2. Detection of Internal and Membrane Virus Antigens in Infected and Transformed Cells.

Cell Type	Anti-Sera						
	CMV[a]		HSV[a]		Zoster[a]		SV40 T[b]
	Internal	Membrane	Internal	Membrane	Internal	Membrane	
CMV transformed	+	+	−	−	−	−	−
CMV-tumor cells	−	+	−	−	−	−	−
CMV-infected[c]	+	+	−	±	−	−	−
HSV-2 infected[c]	−	±	+	+	−	−	−
Human embryo lung	−	−	−	−	−	−	−
Hamster embryo fibroblast	−	−	−	−	−	−	−
PARA[d]	−	−	−	−	−	−	+

[a] Human convalescent sera.
[b] Sera from hamsters bearing SV40 induced tumors.
[c] Human embryonic lung (HEL) and hamster embryonic fibroblasts (HEF) infected with CMV (AD-169) or HSV respectively.
[d] PARA transformed HEF cells.

44

in cell transformation is unknown. We had reported that when cells were pretreated with 5-iodo-2'-deoxyuridine (IUdR) and infected with CMV there appeared to be an enhancement of virus replication (54). In fact, non-permissive cells could be rendered permissive for virus replication using this procedure (55). In addition, in cells that had been pretreated with IUdR and infected with CMV, there was increased uptake of ^{3}H-thymidine (^{3}H-TdR) by infected cells. To determine if this was due to increased cell DNA synthesis or to virus DNA synthesis, we extracted the DNA from these cultures using sarcosyl and pronase and then centrifuged the DNA preparations to equilibrium in CsCl. Fractions of the gradient were collected and the amount of ^{3}H-TdR incorporated into acid insoluble material determined.

Figure 3 represents a profile of a CsCl gradient from a 24-48 hour labelling period. Panel A represents cultures pretreated with IUdR and panel B represents the untreated cultures. Figure 4 represents the kinetics of incorporation of ^{3}H-TdR into cell DNA in cells arrested with IUdR and infected with virus or sham infected with growth medium. Virus and cell DNA were separated in isopynic CsCl gradients and the total counts of ^{3}H-TdR incorporated into cell DNA for each consecutive 24 hour period determined. There was an increase in the uptake of ^{3}H-TdR beginning at 24 hours post-infection in the cultures arrested with IUdR. In a similar study with cells which had not been arrested with IUdR, the results obtained were somewhat different. In these cells, there was little difference between the rate of DNA synthesis in infected and uninfected cells.

These results prompted us to consider whether or not we were dealing with a phenomenon induced by IUdR pretreatment or that DNA synthesis was in fact induced by CMV. Another method of arresting cells was used to determine if similar results would be obtained. The stimulation of cell DNA synthesis in cells arrested using low serum concentrations and infected with either SV40, polyoma or adenoviruses has been reported by many investigators. We attempted to use this procedure in our system. Monolayers of human embryonic lung cells were exposed to 0.2% serum for 48 hours. Cells were either infected with CMV or sham infected and pulsed for consecutive 24 hour periods. The data from this study indicated that CMV increased uptake of ^{3}H-TdR in serum arrested cells.

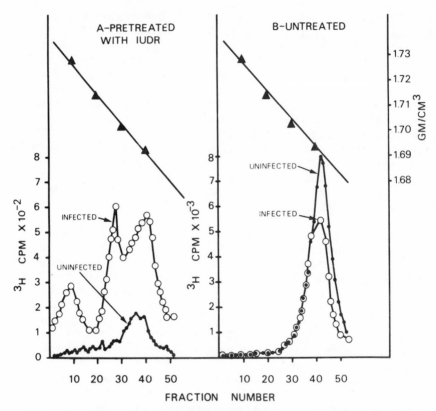

Fig. 3. Neutral isopycnic centrifugation of DNA extracted from CMV-infected and uninfected HEL cells. A, cultures
pretreated with IUdR; B, untreated cultures. Replicating
cell cultures were exposed to either normal Eagle's growth
medium or medium containing 100 μg of IUdR per ml for 96 h
prior to infection. Cultures were infected as described in
the text and then labeled with ^{3}H-Tdr for 24 to 48 h after
infection. DNA was extracted and centrifuged in CsCl gradients (initial density 1.744 g/cm^3) in a Beckman 40.3 rotor
at 30,000 rpm for 60 h at 20°C. Symbols: ○ , counts per
minute from infected cultures; ● , counts per minute from
uninfected cultures ▲ , density determined from refractive
indices. Fraction 9= cell DNA substituted with IUdR, fraction 23= CMV DNA, fraction 41= cell DNA. From St. Jeor
et al. (45).

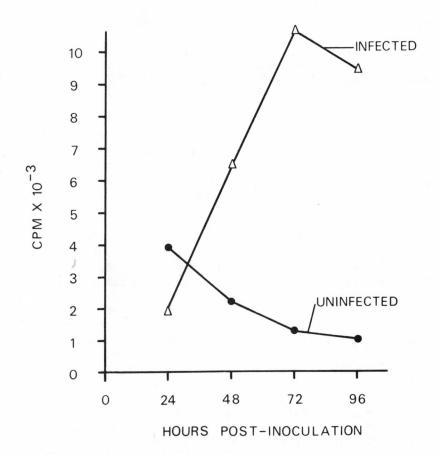

Fig. 4. Kinetics of cell DNA synthesis in HEL cells
pretreated with IUdR and infected with CMV (AD-169) or sham
infected. The cultures were labeled for sequential 24 h
periods with ^{3}H-TdR, and the DNA was analyzed (see legend,
Fig. 3). Data points are plotted at times corresponding to
the end of that particular labeling period and represent the
total counts of ^{3}H-TdR incorporated into DNA with a density
of cell DNA. Symbols: Δ , CMV-infected cultures; ● , uninfected cultures. From St. Jeor et al. (45).

In a continuation of these studies several CMV strains were examined for their ability to stimulate cell DNA synthesis in a variety of cell types. These include rabbit kidney and lung cells, hamster embryo fibroblasts, Vero cells and human embryonic kidney and lung cells. The results of this work are presented in Table 3. Apparently, CMV is able to induce DNA synthesis in several cell types from a variety of animal species.

The ability to stimulate cell DNA synthesis appeared to be a virus coded function. Virus inactivated by either UV light or heat prior to infection produced no stimulation of DNA replication. Furthermore, virus induction of cell DNA synthesis could be blocked by actinomycin D or cycloheximide.

The observed increase in uptake of ^{3}H-TdR in infected cells could have at least two possible explanations. There could be an increase in DNA repair in infected cells or CMV might induce semi-conservative cell DNA synthesis. To determine which of these explanations was responsible for the increased uptake of ^{3}H-TdR the following points were considered. If ^{3}H-TdR was added at the time of infection, repair synthesis would result in the insertion of ^{3}H-TdR into both strands of pre-existing DNA molecules. Semi-conservative synthesis would result in the incorporation of ^{3}H-TdR into newly synthesized DNA. In our studies, cells were treated with IUdR prior to the addition of either CMV or ^{3}H-TdR. Treatment of cells with IUdR results in the presence of DNA molecules containing IUdR incorporated into either one or both strands of a DNA molecule. Consequently, when ^{3}H-TdR is added to virus infected cells, if the increase in ^{3}H-TdR uptake is due to repair synthesis, then ^{3}H-TdR would be incorporated into those DNA strands containing IUdR as well as into unsubstituted strands. Since the IUdR was removed at the time of infection it would no longer be incorporated into cell DNA; consequently, semi-conservative DNA synthesis following infection would result in the incorporation of ^{3}H-TdR into those strands synthesized following infection. In neutral CsCl gradients the type of DNA synthesized could not be determined. It could only be determined that there was an IUdR substituted molecule containing ^{3}H-TdR. However, in alkaline CsCl the two strands of a DNA molecule separate and it can then be determined which strands contain the ^{3}H-TdR. If DNA, having a density similar to cell DNA, contained all of the ^{3}H-TdR then it would indicate that semi-conservative DNA synthesis had been initiated. This would result in a single peak in an alkaline CsCl gradient. If

Table 3. Induction of Mammalian Cell DNA Synthesis by Three Strains of Human Cytomegalovirus.

Cell Lines	Strains of CMV Tested		
	Birch	C-87	AD-169
Human lymphoblastoid	NT[a]	NT	+[b]
Hamster embryo fibroblast	NT	+	+
Rabbit kidney	+	NT	+
Vero	NT	NT	+
Human embryo kidney	+	+	+
Human embryo lung	NT	NT	+

Three strains were tested for their ability to induce cell DNA synthesis in serum or IUdR arrested cells as measured by the incorporation of ^{3}H-TdR into macromolecules with a density of cell DNA.

[a] Not tested.

[b] + indicates at least a 3-fold increase in incorporation of ^{3}H-TdR in infected as compared to control cells.

repair synthesis occurred, ^{3}H–TdR would be found in both IUdR substituted strands (heavy DNA) as well as in unsubstituted DNA (light DNA).

To determine which type of DNA synthesis was occurring, we infected IUdR arrested HEL cells at 12 hours post-infection and the cultures were pulsed with ^{3}H–TdR for an additional 24 hours. At the time the cells were harvested no detectable CMV DNA had been synthesized. The DNA was then analyzed in both alkaline and neutral CsCl gradients. Two bands of DNA were separated in the neutral gradient and a single band was found in the alkaline gradients. This result indicated that CMV had initiated semi-conservative cell DNA synthesis. Using this same experimental design, we found that CMV induced semi-conservative DNA replication in abortive as well as productive infections.

Unpublished observations from our laboratory further indicate that the replication of CMV might be related to the cell cycle. Cells arrested in G-1 with a double thymidine block appear to develop virus cytopathology following infection with CMV more rapidly than do unarrested and infected control cells. Furthermore, our studies indicate that human CMV induces higher levels of cellular thymidine kinase rather than inducing a new virus thymidine kinase following infection.

Partial expression of a virus genome can occur in a cell as the result of infection with a defective virus particle. If the genetic material expressed includes a gene concerned with production of a product which stimulates cell replication, it might be produced in abnormal amounts because controlling factors were not present as a result of the partial expression of the virus genome. If this cell were to survive and develop an abnormal growth pattern the end result could be a malignant cell. Thus, malignancy would be the result of the expression of a virus gene which had evolved to perform a function in virus replication rather than to malignantly transform cells.

Whether or not cytomegalovirus or herpes simplex virus are important in human malignancy remains unknown. However, there is little doubt that they are capable of inducing cell transformation, often resulting in a malignant phenotype.

ACKNOWLEDGEMENTS

This study was conducted under Contract No. 70-7024 within the Virus Cancer Program of the National Cancer Institute, National Institutes of Health, U.S. Public Health Service.

These studies reported in this article were made possible by the continued collaboration efforts of Dr. Ronald Duff, Dr. Thomas Albrecht, and Ms. Jean Li. The technical support of Mr. Myron Katz is a valuable component of this study.

REFERENCES

1. F. Rapp and M.A. Jerkofsky, In "Cancer: A Comprehensive Treatise" (ed. Frederick F. Becker) Plenum Press, New York, in press, (1974).
2. J.S. Butel, S.S. Tevethia and J.L. Melnick, Adv. in Cancer Res. 15,1 (1972).
3. H.M. Temin, Proc. Nat. Acad. Sci. 69,1016 (1972).
4. K. Nazerian, J.J. Solomon, R. L. Witter and B.R. Burmester, Proc. Soc. Exp. Biol. Med. 127,177 (1968).
5. A.E. Churchill and P.M. Biggs, Nature(London) 215,528 (1967).
6. A.E. Churchill, L.N. Payne and R.C. Chubb, Nature (London) 221,744 (1969).
7. B. Lucké, J. Exp. Med. 68,457 (1938).
8. D.W. Fawcett, J. Biophys. Biochem. Cytol. 2,725 (1956).
9. P.D. Lunger, Virology 24,138 (1964).
10. K.A. Rafferty, Jr., Ann. N.Y. Acad. Sci. 126,3 (1965).
11. A. Granoff, In "Oncogenesis and Herpesviruses" (P.M. Biggs, G. de-Thé, and L.N. Payne, eds), pp.171-182 (1973).
12. L.V. Meléndez, R.D. Hunt, M.D. Daniel, F.G. Garcia and C.E.O. Fraser, Lab. Anim. Care 19,378 (1969).
13. L.V. Meléndez, R.D. Hunt, N.W. King, H.H. Barahona, M.D. Daniel, C.E.O. Fraser and F.G. Garcia, Nature (New Biol.) 235,182 (1972).
14. H.C. Hinze, Int. J. Cancer 8,514 (1971).
15. G.D. Hsiung and L.S. Kaplow, J. Virol. 3,355 (1969).
16. M.J. van der Maaten and A.D. Boothe, Archiv für die Gesamte Virusforsch. 37,85 (1972).

17. M.A. Epstein, B.G. Achong and Y.M. Barr, Lancet 1,702 (1964).
18. G. Klein, In "The Herpesviruses" (A.S. Kaplan, ed) The Epstein-Barr Virus (EB), New York, Academic Press, in press, (1974).
19. Z.M. Naib, A.J. Nahmias and W.E. Josey, Cancer 19,1026 (1966).
20. W.E. Rawls, W.A.F. Tompkins and J.L. Melnick, Am. J. Epid. 89,547 (1969).
21. E. Adam, R.H. Kaufman, J.L. Melnick, A.H. Levy and W.E. Rawls, Am. J. Epid. 96,427 (1972).
22. L. Aurelian, Fed. Proc. 31,1651 (1972).
23. N. Frenkel, B. Roizman, E. Cassai and A. Nahmias, Proc. Nat. Acad. Sci. 69,3784 (1972).
24. A.C. Hollinshead and G. Tarro, Science 179,698 (1973).
25. A.B. Sabin and G. Tarro, Proc. Nat. Acad. Sci. 70,3225 (1973).
26. A. Buchan and D.H. Watson, J. Gen. Virol. 4,461 (1969).
27. T. Ogino and F. Rapp, Virology 46,953 (1971).
28. D.B. Davis, W. Munyon, R. Buchsbaum and R. Chawda, J. Virol. 13,140 (1974).
29. A.J. Nahmias, Z.M. Naib, W.E. Josey, F.A. Murphy and C.F. Luce, Proc. Soc. Exp. Biol. Med. 134,1065 (1970).
30. V. Defendi and F. Jensen, Science 157,703 (1967).
31. R. Duff, P. Knight and F. Rapp, Virology 47,849 (1972).
32. R. Duff and F. Rapp, J. Virol. 8,469 (1971).
33. W. Collard, H. Thornton and M. Green, Nature(New Biol.) 243,264 (1973).
34. R. Duff and F. Rapp, J. Virol. 12,209 (1973).
35. C.W. Hiatt, E. Kaufman, J.J. Helprin and S. Baron, J. Immunol. 84,480 (1960).
36. C. Moore, C. Wallis, J.L. Melnick and M.D. Kuns, Infect. and Immunol. 5,169 (1972).
37. F. Rapp, J. Li and M. Jerkofsky, Virology 55,339 (1973).
38. L. Kutinová, V. Vonka and J. Brouvek, J. Nat. Cancer Inst. 50,759 (1973).
39. G. Darai and K. Munk, Nature(New Biol.) 241,268 (1973).
40. C.P. Duvall, A.R. Casazza, P.M. Grimley, P.P. Carbone and W.P. Rowe, Ann. Intern. Med. 64,531 (1966).
41. P.G. Dyment, S.J. Orlando, H. Isaacs, Jr. and H.T. Wright, Jr., J. Pediat. 72,533 (1968).
42. P. Diosi, E. Moldovan and N. Tomescu, Brit. Med. J. 4, 660 (1969).
43. E. Gönczöl and L. Váczi, J. Gen. Virol. 18,143 (1973).

44. D.J. Lang, L. Montagnier and R. Latarjet, Soc. for Ped. Res., Atlantic City, N.J. p.70 (1970).
45. S.C. St. Jeor, T.B. Albrecht, F.D. Funk and F. Rapp, J. Virol. 13,353 (1974).
46. F. Rapp, L.E. Rasmussen and M. Benyesh-Melnick, J. Immun. 91,709 (1963).
47. A. Fioretti, T. Furukawa, D. Santoli and S.A. Plotkin, J. Virol. 11,998 (1973).
48. T. Albrecht and F. Rapp, Virology 55,53 (1973).
49. R. Dulbecco, L.H. Hartwell and M. Vogt, Proc. Nat. Acad. Sci. 53,403 (1965).
50. P. Gerber and B.H. Hoyer, Nature(New Biol.) 231,46 (1971).
51. D. Gershon, L. Sachs and E. Winocour, Proc. Nat. Acad. Sci. 56,918 (1966).
52. M. Takahashi, T. Ogino, K. Baba and M. Onaka, Virology 37,513 (1969).
53. M. Takahashi, G.L. van Hoosier, Jr. and J.J. Trentin, Proc. Soc. Exp. Biol. Med. 122,740 (1966).
54. S. St. Jeor and F. Rapp, J. Virol. 11,986 (1973).
55. S. St. Jeor and F. Rapp, Science 181,1060 (1973).

SV40 GENE FUNCTION

Peter Tegtmeyer
Case Western Reserve University

Tumor viruses have provided the molecular biologist
with a tool to ask specific and relevant questions about
cellular control mechanisms and the nature of the neoplastic
process. SV40, a small simian virus containing only enough
DNA to code for four to eight proteins, stimulates the
growth of certain host cells in a dramatic and reproducible
way. This interaction of the virus and its host cell requires a high degree of cooperation between viral and cellular genes. Thus a detailed analysis of the function of
only a few viral genes may eventually elucidate the mechanism by which the growth of cells is controlled. A number
of laboratories have undertaken this study through the
use of viral mutants.

SV40 may cause either productive infection of permissive monkey kidney cells or transforming infection of
restrictive mouse cells. Other semi-permissive cells,
including human and rabbit cells, may undergo either productive or transforming infection. The early events in each
mode of infection are quite similar in many respects. Stable
early RNA is transcribed from 30-50% of the early strand
of viral DNA (15,20,31). Shortly thereafter the virus-
specific T antigen (1,24) and U antigen (19) appear in
infected nuclei, and the synthesis of cellular DNA (10,11)
and protein (17) is stimulated in most host cells.

In productive infection, the expression of an early
viral gene initiates the replication of viral DNA (34).
Stable late RNA is transcribed from 50-70% of the late
strand of viral DNA (15,20,31) and is translated to produce
three capsid proteins (6) with molecular weights of approximately 46,000 (VP1), 39,000 (VP2), and 28,000 (VP3) daltons.
The viral DNA and associated cellular histones are assem-
bled with capsid proteins into infectious virions (22).
During the course of productive infection, viral DNA may
be integrated into cellular DNA (12), but the significance
of integration in the lytic cycle remains to be determined.

In transforming infection, viral DNA replication

and late genome expression cannot be detected. The infected cells escape growth control transiently in abortive transformation or permanently in stable transformation (33). Stably transformed cells contain integrated viral DNA (8) and continuously synthesize early viral RNA (15,31) and proteins (1,19). The persistence of viral proteins in stably transformed cells suggests that the continued expression of viral genes may influence the properties of the transformed cells. Abortive transformation could be the result of a failure to initiate the integration process or the reversal of unstable integration. Integration must be reversible, since the fusion of transformed cells with uninfected permissive cells regularly results in productive infection (9).

Because of the small size of the SV40 genome, the viral genes required for transformation have been assumed to be also essential for productive infection. This assumption has been the basis for the selection of mutants for the genetic analysis of the transformation process. Temperature-sensitive (ts) mutants are particularly well suited to functional studies because the gene containing the mutation continues to function at the permissive temperature (33°C in these studies) but fails to function at the restrictive temperature (39-41°C). The study of ts mutants of SV40 in permissive cells may define the specific biochemical functions of each viral gene. The parallel study of the same mutants in semi-permissive or restrictive cells should indicate which of the viral genes are required to initiate or maintain the transformed state. The requirement of a ts gene product for the initiation of a function can be determined by comparing viral functions during continuous incubation at the permissive and restrictive temperatures. The requirement of the same gene product for the maintenance of a function can be determined by initiating the function at the permissive temperature and subsequently shifting to the restrictive temperature while monitoring that function. Furthermore, the most significant functional information is obtained during the course of a temperature shift. Intermediates produced before the introduction of a sudden temperature block accumulate at the ts step thereby pinpointing the location of that step. Temperature-sensitive mutants are the only class of mutant which allows this sort of instant genetic manipulation within a given infected cell.

MUTANT GROUPS

Four complementation groups of SV40 ts mutants have been identified (2,5,18,27,36). Groups B and C are defective in late functions in productive infection and cannot be distinguished from wild-type (WT) virus in transforming infection.

The group D mutants include ts101 (27) and D202 (2). When permissive cells are infected with virions, these mutants are defective in all known early and late functions at the restrictive temperature. When cells are infected with mutant DNA rather than virions, productive infection is not defective, yet the resulting progeny virions are still temperature-sensitive (27). Infection of restrictive cells by ts101 results in reduced abortive and stable transformation at both the permissive and restrictive temperatures (28). Cells transformed by ts101 at the permissive temperature maintain the transformed phenotype at the restrictive temperature (Robb, personal communication). The simplest interpretation of these findings is that a virion protein of the mutant is altered in such a way, that viral DNA is not sufficiently uncoated to allow early gene expression under restrictive conditions. However, the immunological expression of T antigen in cells transformed by the mutant has been reported to be partially temperature-sensitive (26). Clearly the D function has not yet been completely defined. It will be important to know whether the D mutations map in the early region of the viral genome.

Group A mutants are defective in an early function in productive infection and fail to initiate the transformation of restrictive cells at 39–41°C (34). This group is especially well suited for physiological characterization, since the ts defect blocks viral DNA replication, transcription of late RNA, and the synthesis of late protein. The role of this ts viral gene (gene A) in the maintenance of the transformed phenotype has not been completely determined. The function of SV40 gene A will be described in the following sections.

LEAKINESS

Temperature-sensitive mutants originate from point mutations which alter the amino acid composition of particular gene products so that a temperature-dependent change in conformation results in decreased function at

57

the restrictive temperature. Any residual activity is
known as leakiness. A knowledge of the extent of leakiness
of each mutant is crucial to interpreting the results of
experiments with the mutants. If a mutant is blocked in
a particular function, the altered gene clearly must be
directly or indirectly required for that function. However,
the failure of a mutation to significantly inhibit a given
function cannot rigorously exclude the requirement of the
altered gene for expression of that function if the mutant
is leaky.

Table 1 compares the function of three A mutants to
that of WT virus under permissive and restrictive conditions. Permissive cells were infected with virions, and
the yield of infectious DNA was measured after a single
cycle of growth. Mutants A28 and A30 produced little if
any infectious DNA at 41°C but were very leaky at 39°C.
Mutant A58 did not function at either temperature.

When permissive cells were inoculated with A mutant
DNA rather than virions, infection was also temperature-
sensitive (Table 2). The input DNA resulted in no background infectivity when the single cycle yield of virions
was subsequently assayed under conditions in which viral
DNA is not infectious. Mutant A58 again failed to produce
any detectable progeny virus at 39-41°C under these circumstances. Furthermore, these results exclude the possibility that a block in adsorption, penetration, or uncoating
of the viral genome causes the defect in infection by the
A mutants. In contrast, the extracted DNA of mutant D202
was as infectious as WT DNA.

VIRAL DNA REPLICATION

The replication of SV40 DNA has been well characterized. Initiation of replication of the closed circular
viral DNA starts at a specific site within the late region
of the viral DNA (7,16,21). Propagation of the daughter
strands proceeds bidirectionally and symmetrically (7).
During the process of replication the superhelical parental
DNA unwinds progressively (14,30,32). The newly replicated
branches are relaxed and the unreplicated parental strands
remain superhelical in most of the replicating molecules.
Segregation of the daughter molecules requires a final
nick in at least one of the parental DNA strands, and the
progeny DNA must be converted to the superhelical configuration to complete the cycle of replication. Thus each round
of replication can be divided into at least four distinct
stages: initiation, propagation, segregation, and com-

TABLE 1.   YIELD OF DNA FROM ONE CYCLE OF GROWTH AFTER
INFECTION BY VIRIONS

Virus	DNA Yield (PFU/ml)		
	33°C	39°C	41°C
WT	1×10^5	2×10^5	1×10^5
A28	9×10^4	1×10^4	$< 10^1$
A30	2×10^5	1×10^4	$< 10^1$
A58	8×10^4	$< 10^1$	$< 10^1$

Permissive CV-1 monkey cells were infected at a
multiplicity of 10 plaque-forming units/cell. After 3
days at 39°C or 41°C and after 5 days at 33°C, viral DNA
was selectively extracted (13) and plaque assayed.

TABLE 2.   YIELD OF VIRIONS FROM ONE CYCLE OF GROWTH
AFTER INFECTION BY EXTRACTED VIRAL DNA

Viral DNA	Virion Yield (PFU/ml)		
	33°C	39°C	41°C
WT	5×10^6	1×10^7	3×10^6
A58	2×10^6	0	0
D202	1×10^6	2×10^6	5×10^5

Permissive CV-1 monkey cells were infected at a
multiplicity of 0.1 plaque-forming units of DNA/cell.
After 3 days at 39°C or 41°C and after 5 days at 33°C,
the cultures were frozen and thawed and plaque assayed
for total virus production.

pletion (Fig. 1). The duration of the entire process lasts 10-15 minutes for the average molecule. Most of this time is required for propagation since the replicative intermediate (RI) is the predominant molecule labeled by a 15 minute pulse with ^{3}H-thymidine.

The process of replication can be monitored by electrophoresis of selectively-extracted viral DNA through SDS-agarose gels (Fig. 2). Closed circular DNA I migrates rapidly through the gel because of its small molecular weight and compact shape. Nicked circular DNA II has the same molecular weight but a relaxed conformation and thus moves more slowly. Early replicative intermediates migrate between DNA I and II, and late RI molecules migrate more slowly than DNA II, again reflecting differences in the size and shape of the molecules (35).

Mutant A58 fails to replicate viral DNA during continuous infection at 41°C. In order to localize the defect in the replication cycle of viral DNA, DNA synthesis was allowed to reach maximum levels at 33°C and then was exposed to a sudden temperature block. Immediately after the shift to 41°C, the infected cells were labeled for 15 minutes with ^{3}H-thymidine. The labeled viral DNA was extracted and analyzed by gel electrophoresis. The results (Fig. 3) fulfilled the predictions for a mutant blocked exclusively in the initiation of the DNA cycle. DNA synthesis continued, but at a reduced rate for approximately 15 minutes, the length of a single round of viral replication. Most of the mutant DNA which had been labeled during the period of decreasing synthesis was converted into mature DNA I in contrast to labeled WT DNA which was predominantly in the RI form under the same conditions. The findings suggest that the rounds of mutant DNA replication initiated before the temperature block were completed, but little, if any, initiation occurred subsequent to the block.

Pulse-chase experiments further confirmed that mutant RI molecules initiated at 33°C were propagated, segregated, and converted into mature DNA I at 41°C at the same rate as in infection by WT virus (Fig. 4). Infected cells were pulse-labeled with ^{3}H-thymidine for 15 minutes at 33°C, shifted to 41°C, and chased with excess unlabeled thymidine for increasing intervals before extraction and analysis. The mutant replicative intermediates were converted into DNA I at the same rate as WT molecules, even though the total new synthesis of mutant DNA was significantly reduced under identical conditions.

D mutants also fail to replicate viral DNA during continuous infection at 41°C. However, a shift from the

60

SV40 DNA REPLICATION

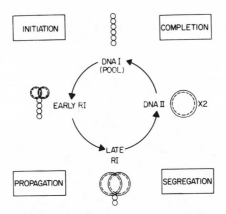

Fig. 1. The cycle of SV40 DNA replication showing stages at which the cycle could be blocked.

GEL ELECTROPHORESIS OF SV40 DNA

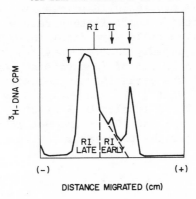

Fig. 2. Diagrammatic representation of the analysis of SV40 molecules by electrophoresis through 1.5% agarose gels (35).

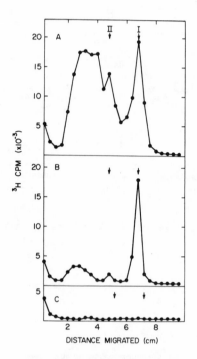

Fig. 3.   DNA synthesis in permissive infection by WT or mutant A58 after a shift in temperature from 33°C to 41°C.   The infected cells were first incubated for 72 hr at 33°C and then rapidly shifted to 41°C.   The cultures were labeled with 50 µCi of ^{3}H-thymidine per ml of medium for 15 minutes.   The viral DNA was selectively extracted by the method of Hirt (13)   and analyzed by gel electro-phoresis.   (A) WT DNA labeled 0-15 minutes after the tem-perature shift.   (B) and (C) Mutant DNA labeled 0-15 and 30-45 minutes after the temperature shift.

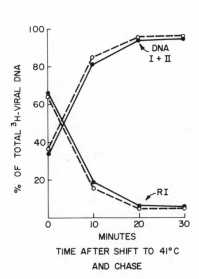

Fig. 4. The processing of replicative intermediate molecules of SV40 DNA after a shift in temperature from 33°C to 41°C. Permissive cells infected by WT virus or mutant A58 were incubated at 33°C for 72 hr and were pulse-labeled with 50 µCi of ^{3}H-thymidine per ml of medium for 15 minutes. Immediately thereafter, the cultures were rapidly shifted to 41°C and the isotope was chased with excess unlabeled thymidine for the periods of time shown above. At the end of the chase period, viral DNA was selectively extracted and analyzed by gel electrophoresis. The relative quantity of label in the various molecular forms of DNA is plotted above, WT (●—●) or mutant (0--0). DNA II never constituted more than 5% of the label in a given sample.

permissive to the restrictive temperature has no blocking effect (3). Thus gene A is the only viral gene identified to date, which is directly required for viral DNA synthesis. Presumably, the cell provides most if not all of the remaining proteins required to propagate and complete viral DNA. Clearly, the viral initiation system must effectively interact with the cellular replicating machinery.

CELLULAR DNA REPLICATION

The evidence that SV40 gene A codes for a specific initiator of viral DNA synthesis suggested the possibility that the A protein may also interact with cellular DNA to induce host DNA synthesis. Table 3 compares the induction of DNA replication in cells infected by WT or mutant virus under permissive and restrictive conditions. WT virus stimulated DNA synthesis in permissive cells approximately tenfold at either 33°C or 41°C. A late mutant, B4, resembled WT virus but induced somewhat less DNA synthesis at 41°C than at 33°C. The most probable explanation of this finding is that WT virus may have begun a second cycle of virus growth and cell stimulation during the course of the experiment. Early mutants A28 and A30 induced less cell DNA synthesis at 41°C than at 33°C, nevertheless the levels of ³H-thymidine incorporation into cellular DNA were well above those of mock-infected cells. These studies suggest that gene A may not be required for the stimulation of host DNA. A note of caution in interpretation is in order. A single interaction of gene A with the host DNA or with viral DNA integrated in host DNA could initiate a complete round of host DNA synthesis which would be easily detected in comparison to a single round of viral DNA replication. Thus even minimal leakiness of gene A could result in misleading data. Studies of induction by the "non-leaky" mutant A58 are not yet complete.

VIRAL TRANSCRIPTION

During productive infection two distinct populations of stable viral RNA are produced (15,20,31). Early in infection before the onset of viral DNA synthesis, RNA is transcribed from 30-50% of one strand of viral DNA. Late in infection after the onset of viral DNA synthesis, the original species of RNA continues to be transcribed and in addition RNA is also transcribed from 50-70% of the opposite strand of viral DNA. Because of the temporal

TABLE 3. INDUCTION OF CELLULAR DNA SYNTHESIS BY SV40

	DNA Synthesis (CPM/culture)	
Virus	33°C	41°C
Mock	5,251 ± 14%	7,596 ± 12%
WT	72,063 ± 11%	63,682 ± 18%
B4	92,461 ± 8%	64,481 ± 8%
A28	119,678 ± 12%	60,168 ± 3%
A30	109,609 ± 14%	41,991 ± 7%

Permissive primary monkey kidney cells were infected at a multiplicity of 50 plaque-forming units/cell. The cultures were labeled with 1 µCi/ml of ^{3}H-thymidine from 24-48 hr after infection at 41°C and from 48-96 hr at 33°C. Cell DNA was selectively extracted and acid-insoluble counts were quantitated.

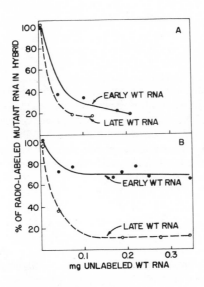

Figure 5. See legend on pg. 66.

TABLE 4. ANTIGENIC PHENOTYPES OF SV40 MUTANTS

Virus	Early Antigens		Late Antigens	
Group	T	U	C	V
D	0	0	0	0
A	↓	0	0	0
B_1	+	0	+	0
B_2	+	+	+	+

Summarized data from Robb et al. as determined by
immunofluorescent assay (29). T antigen accumulation by
A mutants is also decreased by complement fixation assay
(36). C antiserum reacts with VP1 and VP3. At the concentration used, V antiserum reacts only with intact
capsids. Virus groups B_1 and B_2 belong to the same complementation group. B_1 mutants fail to assemble stable
virus particles, whereas B_2 mutants assemble defective
particles.

Fig. 5. RNA-DNA competition-hydrization analysis of
mutant RNA synthesized in permissive cells during continuous
incubation at 41°C or after a temperature shift from 33°C to
41°C. (A) Cells infected by A30 were labeled with 20 μCi/ml
of ^{3}H-uridine 50-70 hr after continuous infection at 41°C.
(B) Cells infected by A30 were incubated at 33°C for 65 hr.
The cultures were rapidly shifted to 41°C and 2.5 hr later
were labeled with 60 μCi/ml of ^{32}P in phosphate-free medium
for 4 hr. RNA was extracted from the infected cultures and
hybridized with viral DNA on nitrocellulose filters in the
presence of increasing amounts of unlabeled early or late
wild-type RNA as described by Cowan (4). Radioactivity
bound to viral DNA in the absence of competing RNA was
5020 cpm in panel A and 725 cpm in panel B.

relationship between viral DNA and RNA synthesis, the role
of gene A in regulating late viral transcription was investigated (4).

The requirement of gene A to initiate early and late
viral transcription was examined during continuous infection
by A30 at 41°C. Fig. 5A illustrates the results of competition-hybridization experiments between radio-labeled mutant
RNA synthesized at 41°C and excess, unlabeled early or
late WT DNA. The binding to viral DNA of mutant RNA,
labeled and isolated late in the course of infection, was
completed almost completely by either early or late WT RNA.
This result indicates that only early RNA was transcribed
in infection by A30. The reverse competition between
labeled early WT RNA and excess unlabeled mutant RNA produced at 41°C was also nearly complete (data not shown).
Thus all of the early sequences but no late sequences were
transcribed during continuous infection by the mutant at
the restrictive temperature.

Temperature-shift experiments were then done to determine whether the synthesis of late mutant RNA required
the continuous expression of gene A. DNA synthesis and
the transcription of late viral RNA were allowed to reach
high levels at 33°C, then gene A expression was blocked
by a shift to 41°C. No viral DNA synthesis could be detected 30 minutes later. Two hours later, RNA in the infected cultures was radio-labeled for 4 hours. Fig. 5B
illustrates the results of competition-hybridization between
the labeled mutant RNA and unlabeled, excess early or late
WT RNA. The mutant showed the same pattern of competition
seen in a parallel temperature-shift experiment that used
WT virus. Competition by an excess of early WT RNA was
limited to 30%, while competition by late WT RNA was almost
complete. Thus the labeled mutant RNA contained sequences
that were transcribed from both early and late genes.
Clearly the synthesis of late viral RNA can occur in the
absence of both concurrent viral DNA synthesis and the
continued expression of gene A. Thus the A function is
directly or indirectly required to initiate, but not to
maintain, synthesis of late RNA during productive infection.

PROTEIN SYNTHESIS

The antigenic phenotypes of three complementation
groups of SV40 (29,36) mutants are summarized in Table 4.
No viral antigens can be detected in cells infected by
ts101. A mutants induce synthesis of T antigen in reduced

quantity but no other antigens can be detected by the im-
muno-fluorescence technique. This finding suggests that
the gene A product is T antigen, since temperature-sensitive proteins usually retain some immunological specificity. If this interpretation is correct, then the presence
of the early U antigen must depend in some way on the synthesis or function of T antigen. The absence of U antigen
can not be easily explained by a block in transcription,
since A mutants transcribe most if not all early RNA sequences. To further complicate the situation, U antigen
cannot be detected in infection by tsBll, a late mutant
blocked in viral assembly (23). These immunological
studies, which measure the accumulation rather than the
synthesis of antigens in infected cells, have been useful
in identifying interesting mutants, but are inadequate
to provide any real insight into the physiology of viral
infection.

Protein synthesis in infected cells was further examined by the electrophoresis of cellular extracts through
SDS, polyacrylamide, slab gels. Fig. 6 shows the gel
patterns of proteins extracted from the nuclei of uninfected
or infected cells. The cells were labeled with ^{35}S-
methionine at 33°C, at 41°C, or after a shift from 33°C
to 41°C. Capsid proteins VP1 and 3 could be easily identified in the nuclear extracts of cells infected by WT
virus at 33°C and 41°C or by A30 at 33°C. In contrast,
capsid proteins could not be detected in continuous infection by A30 at 41°C. When cells infected by A30 for 72
hours at 33°C were shifted to 41°C for 24 hours and then
labeled with ^{35}S-methionine for 1 hour, the capsid proteins
were synthesized at the same rate as in cells infected
by WT virus under the same conditions. These findings
are in complete agreement with the studies showing that
gene A expression is required to initiate but not to maintain continuous late viral transcription.

After the temperature shift, a 72,000 dalton protein
was also identified in the nuclei of cells infected by
A30. This protein has the same apparent molecular weight
as a component synthesized in smaller quantities in uninfected cells or in cells infected by WT virus. In other
experiments (not shown) the 72,000 dalton protein accumulated to a limited extent within the cytoplasm of cells
late in the course of infection by the A mutants after
continuous incubation at 41°C. The simplest interpretation
of these findings is that this protein is a cellular protein
which can be specifically induced by SV40 under certain
conditions of restrictive infection. Nevertheless, it

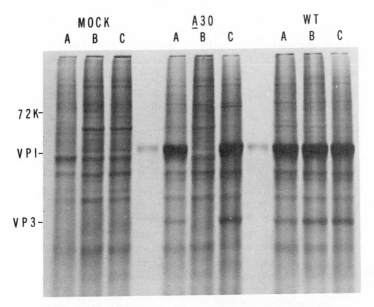

MOCK   A30   WT
A   B   C   A   B   C   A   B   C

72K-

VPI-

VP3-

Fig. 6. Gel patterns of intranuclear proteins synthesized in mock-infected cells, cells infected by A30, or
cells infected by WT virus. The cultures were labeled with
25 µCi/ml of ^{35}S-methionine in methionine-free medium 70-
72 hr after infection at 33°C (A), 47-48 hr after infection
at 41°C (B), or 23-24 hr after a shift from 33°C (72 hr pre-
incubation) to 41°C (C).

TABLE 5. INITIATION OF TRANSFORMATION OF 3T3 CELLS BY SV40

Virus	Transformed Colonies	
	33°C	39°C
Mock	0	0
WT	344	331
B4	368	358
A28	311	2
A30	237	2

Restrictive mouse 3T3 cell monolayers were infected at
a multiplicity of 100 plaque-forming units/cell. Cultures
were diluted 10^{-2} and plated at 33°C and 39°C. Transformed
colonies were counted after 10 days at 39°C and after 20
days at 33°C.

would be premature to exclude the possibility that this
protein is coded by the viral genome. In either case, it
may play an important role in the early viral interaction
with its host cell.

TRANSFORMATION

Studies of mutant gene function in the transformation of restrictive cells present special problems in interpretation, since a temperature-sensitive change in the
mutant viral protein could theoretically be suppressed
by host factors. Indeed, if the mutant protein has more
than one interaction with the host cell, one interaction
could conceivably remain temperature-sensitive while another
would not be in the new host. To avoid these difficulties,
a significant number of representatives of each mutant
group should be examined in more than one restrictive host.
The group $\underline{A}$ mutants are defective in the initiation
of stable transformation of the restrictive mouse 3T3 cell
line, as determined by colony assay under soft agar at
39°C (Table 5). However, when 3T3 cells are transformed
at 33°C and then shifted to 39°C, the cultures maintain
their transformed phenotype and single cells form transformed colonies as well as cells transformed by WT virus.
To verify this finding in another host, semi-permissive
rabbit cells were transformed by three different mutants
in the $\underline{A}$ group. Rabbit cells were chosen because non-trans-
formed cells have a very limited life span in culture and
can be subcultured only 5-10 times even at high cell concentrations. In contrast, rabbit cells transformed by
SV40 can be subcultured indefinitely at 41°C even at very
low cell concentrations. The cells were infected at 33°C
by high multiplicities of virus, so that many cells in each
culture were transformed after the initial exposure to
virus. The cultures were passed until all of the cells
appeared transformed and contained T antigen. Cells grown
at 33°C were dispersed with trypsin and serial, ten-fold
dilutions of the cells were plated in duplicate cultures
at 33°C or 41°C. After an appropriate time, the colonies
were fixed, stained, and counted. The results from one
experiment are presented in Table 6. WT, $\underline{A}$30 and $\underline{A}$58 transformed cells produced transformed colonies with approxi-
mately the same efficiency at both temperatures. $\underline{A}$28 cells
had a significantly reduced plating efficiency at 41°C.
Nevertheless the colonies which grew at higher cell concentrations at 41°C had the phenotype characteristic of transformed cells. Similar results have been obtained with

70

TABLE 6. COLONY FORMATION BY RK CELLS AT 41°C
AFTER TRANSFORMATION AT 33°C

Cell Dilution	WT		A28		A30		A58	
	33°C	41°C	33°C	41°C	33°C	41°C	33°C	41°C
10^{-2}	TM	TM	TM	128	TM	TM	TM	TM
10^{-3}	TM	TM	TM	6	TM	TM	TM	TM
10^{-4}	201	187	105	0	145	143	59	53

Semi-permissive rabbit cells were inoculated with 100
plaque-forming units per cell. The cells were subcultured
until uniformly transformed at 33°C. The cultures were
trypsinized, serially diluted, and plated at 33°C and 41°C.
Colonies were counted after 2 weeks at 41°C and after 3
weeks at 33°C.

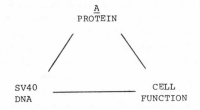

A
PROTEIN

SV40
DNA

CELL
FUNCTION

1. Initiate every
   round of viral
   DNA replication

2. Continuous late
   Transcription

3. Initiate stable
   Transformation

4. Integration?

Fig. 7. Model to explain SV40 gene A expression. The
A protein interacts with SV40 DNA in a specific way, but
the outcome of this interaction is determined by a variety
of cellular functions.

two other rabbit kidney lines independently transformed by A28.

The preponderance of evidence clearly suggests that SV40 gene A is not required to maintain the transformed state. However, the temperature-sensitive plating efficiency of rabbit cells transformed by A28 suggests that SV40 gene A may influence the behavior of transformed cells under some circumstances.

CONCLUSION

The A protein of SV40 clearly plays a central role in productive infection. Complementation studies suggest that gene A codes for a diffusable gene product (34). Its function is not required for early transcription or T antigen synthesis, although T antigen is produced in somewhat reduced quantities as determined by immuno-assay. No other antigens, including the early U antigen, can be detected in cells continuously infected by A mutants at the restrictive temperature. Gene A is directly and continuously required to initiate, but not to propagate or complete, each round of viral DNA replication. Under usual circumstances, the cell cannot provide a host initiator to substitute for the A function in initiating the replication of free viral DNA. Expression of this gene is only transiently required for late transcription and the synthesis of capsid proteins. Significantly, gene A is also required to initiate but apparently not to maintain the stable transformation of restrictive cells. The specific function of the A protein in initiating the transformation process has not yet been determined, but it may be required to establish stable integration.

Fig. 7 presents a model to explain the multiple functions of gene A on the basis of a single molecular activity. The A protein would interact with SV40 DNA in a specific way, for example as an endonuclease, to alter the structure of the viral DNA. This alteration could result in an interaction with a number of different cell proteins to initiate replication, transcription, or perhaps integration. The original structure of viral DNA would be restored at completion of the DNA replication cycle, thereby requiring another interaction with the A protein for continued DNA replication. In contrast, the DNA in transcription complexes or DNA in the integrated state would not usually be restored to the DNA I structure, and the A function would not be required for continued transcription or for maintaining the integrated state.

72

Our present understanding of SV40 gene function is inadequate to explain the induction of cellular DNA synthesis. The ts mutants in complementation groups $\underline{A}$, $\underline{B}$, and $\underline{C}$, which have been examined to date, stimulate the cell under restrictive conditions. Although the virions of $\underline{D}$ mutants fail to induce cellular DNA synthesis, a defect in uncoating of the viral genome would block all viral functions in an indirect way. The apparent failure to isolate mutants blocked directly in the induction process may be explained in a number of ways. The present techniques for identifying ts-mutants have been based on the assumption that all viral genes are essential for productive infection. Yet the induction of host DNA synthesis may not be essential to viral replication, since productive infection proceeds efficiently in the BSC-1 line which is "noninducible" (25). Alternatively, the nature or the size of the inducer gene could make it resistant to temperature-sensitive mutation. The isolation of a mutant with a ts inducer is crucial since its continued function may be required to sustain the transformed state of cells.

ACKNOWLEDGMENTS

These studies were supported by grants VC-98A and PRA-113 from the American Cancer Society.
I would like to acknowledge the importance of the studies of M. Fried, R. Dulbecco, and W. Eckhart on the genetics of polyoma virus to the work reported here and the close cooperation of R.M. Martin, J.A. Robb, J.Y. Chou, K. Cowan, D.D. Anthony, and H.L. Ozer in the isolation and characterization of SV40 mutants.

REFERENCES

1. Black, P.H., W.P. Rowe, H.C. Turner and R.J. Huebner, Proc. Nat. Acad. Sci. U.S.A. 50, 1148 (1963).
2. Chou, J.Y. and R.G. Martin, J. Virol. In Press (1974).
3. Chou, J.Y., J. Avila and R.G. Martin, J. Virol. In Press (1974).
4. Cowan, K., P. Tegtmeyer and D.D. Anthony, Proc. Nat. Acad. Sci. U.S.A. 70, 1927 (1973).
5. Dubbs, D.R., M. Rachmeler and S. Kit, Virology 57, 161 (1974).
6. Estes, M.K., E.S. Huang and J.S. Pagano, J. Virol. 7, 635 (1971).
7. Fareed, G.C., C.F. Garon and N.P. Salzman, J. Virol. 10, 484 (1972).

8. Gelb, L.D., D.E. Kohne and M.A. Martin, J. Mol. Biol. 57, 129 (1971).
9. Gerber, P., Virology 28, 501 (1966).
10. Gershon, D., L. Sachs and E. Winocour, Proc. Nat. Acad. Sci. U.S.A. 56, 918 (1966).
11. Hatanaka, M. and R. Dulbecco, Proc. Nat. Acad. Sci. U.S.A. 56, 736 (1966).
12. Hirai, K. and V. Defendi, J. Virol. 9, 705 (1972).
13. Hirt, B., J. Mol. Biol. 26, 365 (1967).
14. Jaenisch, R., A. Mayor and A.J. Levine, Nature New Biol. 233, 72 (1971).
15. Khoury, G., J.C. Byrne and M.A. Martin, Proc. Nat. Acad. Sci. U.S.A. 69, 1925 (1972).
16. Khoury, G., M.A. Martin, T.N.H. Lee, K.J. Danna and D. Nathans, J. Mol. Biol. 78, 377 (1973).
17. Kiehn, E.D., Virology 56, 313 (1973).
18. Kimura, G. and R. Dulbecco, Virology 49, 394 (1972).
19. Lewis, A.M., Jr. and W.P. Rowe, J. Virol. 7, 189 (1971).
20. Lindstrom, D.M. and R. Dulbecco, Proc. Nat. Acad. Sci. U.S.A. 69, 1517 (1972).
21. Nathans, D. and K.J. Danna, Nature New Biol. 236, 200 (1972).
22. Ozer, H.L., J. Virol. 9, 41 (1972).
23. Ozer, H.L. and P. Tegtmeyer, J. Virol. 9, 52 (1972).
24. Rapp, F., T. Kitahara, J.S. Butel and J.L. Melnick, Proc. Nat. Acad. Sci. U.S.A. 52, 1138 (1964).
25. Ritzi, E. and A.J. Levine, J. Virol. 5, 686 (1970).
26. Robb, J.A., J. Virol. 12, 1187 (1973).
27. Robb, J.A. and R.G. Martin, J. Virol. 9, 956 (1972).
28. Robb, J.A., H.S. Smith and C.D. Scher, J. Virol. 9, 969 (1972).
29. Robb, J.A., P. Tegtmeyer, A. Ishikawa and H.L. Ozer, J. Virol. 13, 662 (1974).
30. Salzman, N.P., E.D. Sebring and M. Radonovich, J. Virol. 13, 662 (1974).
31. Sambrook, J., P.A. Sharp and W. Keller, J. Mol. Biol. 70, 57 (1972).
32. Sebring, E.D., T.J. Kelly, Jr., M.M. Thoren and N.P. Salzman, J. Virol. 8, 478 (1971).
33. Smith, H.S., C.D. Scher, and G.J. Todaro, Virology 44, 359 (1971).
34. Tegtmeyer, P., J. Virol. 10, 591 (1972).
35. Tegtmeyer, P. and F. Macasaet, J. Virol. 10, 599 (1972).
36. Tegtmeyer, P. and H.L. Ozer, J. Virol. 8, 516 (1971).

TEMPERATURE SENSITIVE MUTATIONS: TOOLS FOR UNDERSTANDING GROWTH REGULATION IN SOMATIC ANIMAL CELLS

Claudio Basilico, Stuart J. Burstin,
Daniela Toniolo and Harriet K. Meiss

Department of Pathology
New York University School of Medicine
New York, N.Y. 10016

The study of the phenomena following infection of cells by tumor viruses, that lead to the establishment of malignancy has led to a greater understanding of the mechanisms of carcinogenesis. However, the basic differences between normal and transformed cells are still very elusive. One of the reasons for this lack of knowledge is that we still know very little about growth regulation in normal cells. Genetic tools should be helpful in this type of studies with somatic animal cells, as they have been extremely important in the study of bacterial and viral physiology. They should enable us to obtain valuable information on the type of changes which characterize a transformed cell and on the regulatory mechanisms which are altered after malignant transformation.

Generally very useful for this type of study are conditional lethal, temperature-sensitive (ts) mutations. These missense mutations generally cause an amino acid substitution in a protein which, as a result, retains its function at low temperature, but loses its functionality at high temperature. Ts mutations have been shown to arise in most organisms studied and to occur all over the genome (1,2), thus allowing the examination of a wide spectrum of functions. Most important, the conditional expression of ts mutations provides an internal control and therefore a unique way of testing directly the involvement of a specific gene-product in determining a specific phenotype.

In the past years we have been concerned with a number of studies involving the use of ts cell mutants. One such study led to the isolation of cells transformed by SV40 virus which expressed the transformed phenotype in a temperature-dependent manner. These cells owe their behavior to a cellular mutation which at high temperature suppresses the expression of transformation (3,4). These findings showed that while the virus is the initiator of a complex chain of events which ultimately lead to malignancy, the host cell can exert various types of control over the expression of the transformed phenotype.

These studies, however, will not be the subject of this article. We wish to describe here a long range approach to the understanding of cell physiology. We undertook the isolation of temperature-sensitive mutants of somatic animal cells, i.e. cells which because of specific mutations are capable of growth at a low permissive temperature but not at a high non-permissive temperature. We hoped that by studying specific cellular functions and the effect of specific defects on cellular growth, we would be able to gain valuable information as to the regulation of growth in animal cells.

Selection and Isolation of Temperature-Sensitive Mutants of BHK 21 Cells.

For our studies we choose BHK 21 cells, a continuous line of hamster fibroblasts (5). The reasons for such a choice were the following: a) These cells have a relatively short generation time. b) They plate with good efficiency at any temperature. c) They have a diploid chromosome number. This is probably of paramount importance since, as shown in many organisms and as it will be seen later also in our system, most ts mutations are recessive. Therefore the expression of such mutations would be very difficult in cells characterized by a hyperdiploid chromosome complement. d) Finally, BHK cells are, insofar as continuous lines can be defined, relatively normal with respect to their growth properties, and they can be transformed in vitro by oncogenic viruses and by chemical carcinogens.

The procedures for the selection and isolation of ts mutants of this line have been already described (6,7). The temperatures chosen were 39° as non-permissive temperature and 33.5° as permissive temperature. We have now isolated over 70 "good" (see later) ts mutants of BHK cells. The frequency at which they were obtained was in the order of 10^{-7}.

General Properties of the ts BHK Mutants.
The mutants we have isolated fall broadly into three categories (6). The first category includes mutants that we have defined "good", as they have a low reversion frequency and low leakiness. The second type of mutants is characterized by a high reversion rate. These mutants were not characterized further. The third type of mutants displayed density-dependent growth. In particular, they gave no colonies when plated at 39° at low cell number, but when plated at high cell density they grew almost like wild-type cells. For obvious reasons these mutants are not very useful when dealing with large quantities of cells and they also have not been characterized any further.

In conclusion, all of our studies have been carried out with the so-called "good" mutants, and although most of them seem to fall into different classes with respect to their functional defect, they have some characteristics in common, which are the following.

Reversion is low ($\leq 1 \times 10^{-6}$), but can be increased by ten to twenty-fold following treatment with mutagens (6,7). Thus our mutations are likely to derive from single base substitutions. The mutants are genetically stable and homogenous, since recloning the cell population yields clones which have essentially the same characteristics of the parental cells.

It was of importance to determine whether these mutations were dominant or recessive. Most ts mutations have been shown to be recessive in viruses, bacteria, fungi or in Drosophila (1,8,9), but in our case we were dealing with diploid cells and therefore it could be thought that the most frequent type of temperature-sensitive mutation obtained would be of a dominant type. Such mutations would still be useful, but might make certain types of experiments, such as complementation analysis, impossible to perform.

In principle, there are at least four ways in which recessive mutations could be expressed in a diploid cell: a) They could represent mutations in genes which are normally present in a haploid state (e.g., on the X-chromosome). b) They could represent mutations in genes that because of a deletion of the homologous locus or chromosome became haploid in the cell in question. c) They represent originally heterozygous mutations that through processes of somatic crossing over have become homozygous in the mutant cell. d) A fourth possibility is that somatic cells may already possess a high degree of heterozygosis; therefore, only a change of the dominant allele should be necessary to change the cells' phenotype.

The technique we used to determine dominance or recessiveness was that of somatic cell hybridization. Mutants to be tested were mixed in the presence of Sendai virus, and after a few days at 33° the cells were plated at 39°. At this temperature the two parental cells cannot grow, while the hybrid cells should be capable of sustained growth if the two mutations were recessive and able to complement. After an appropriate period of time at 39° the plates were stained and the number of colonies arising in the cell population were counted (6). Representative results are shown in Table 1. Most mutant crosses seem to be able to give rise to hybrids capable of growth at 39°, with some exceptions. However, these exceptions included mutants which, although incapable of hybridizing between themselves, were capable of making viable hybrids when crossed with other mutants. Therefore, the simplest interpretation of these results was the following: a) These ts mutations are recessive. b) Most mutants are not identical, since they complement, with the exception of some mutants which therefore belong to the same complementation group. However, since chromosome loss is not infrequent in hybrid cells, and we detect our hybrids by their ability to grow at 39°, it was possible that out of the whole possible hybrid cells we had selected only for the ones that had lost the chromosomes bearing a dominant gene for the ts mutation. Therefore, a more rigorous test for recessiveness or dominance would have been to select for hybrids at 33°, in the absence of selective pressure for ability to grow at 39°, and then test the ability of these hybrids to multiply at a high temperature.

TABLE 1.   Hybridization Among ts Mutants of BHK Cells*

Mutants Cross	Frequency of Colonies at 39°	33° in HAT Medium	Comple-mentation	Hybrids isolated at 33° able to grow at 39°
ts422E x tsBCH	3.7×10^{-4}		+	
ts422E x tsBCL	3.6×10^{-4}		+	
tsBCH x tsBCL	6.0×10^{-6}		-	
ts422E x tsAF8	1.0×10^{-4}		+	
tsT15 x tsT23	2.2×10^{-4}		+	
tsT15 x tsT22	2.0×10^{-4}		+	
tsT15 x tsT60	0.9×10^{4}		+	
tsT23 x tsT60	$<1.1 \times 10^{-6}$		-	
tsT23 x tsT22	2.2×10^{-4}		+	
tsT15 x B1 (ts⁺, TK⁻)	5.2×10^{-4}	4.3×10^{-4}	+	7/7
tsT23 x B1 (ts⁺, TK⁻)	2.5×10^{-4}	3.4×10^{-4}	+	6/6
tsT22 x B1 (ts⁺, TK⁻)	2.8×10^{-4}	3.5×10^{4}	+	8/8

*Hybridization was performed as described (6). After incubation at 33° for 2 days, the cells were replated under the conditions indicated. The parent mutant cell lines were also plated individually under the same conditions. All ts mutants plated individually gave a frequency of colonies at 39° of $<1 \times 10^{-6}$, with the exception of ts BCH (8×10^{-6}). All T mutants are ts and HGPRT-.

†HAT: 10^{-5}M aminopterin, 4×10^{-5}M thymidine, 10^{-4}M hypoxanthine, 10^{-5}M glycine; does not allow growth of TK⁻ cells.

By using ts mutants derived from a BHK cell deficient in the hypoxanthine guanine phosphoribosyl transferase enzyme (HGPRT⁻), it was possible to cross these ts HGPRT⁻ mutants with cells wild-type for the ts mutation (ts^+), although biochemically marked by a thymidine kinase (TK) deficiency, and isolate the hybrids at the permissive temperature. If the ts mutation carried by these cells were dominant, it would have been able to inhibit also the wild-type genome, when tested at 39°.

The results of such an experiment showed that hybrid colonies formed at 33° were all capable of growth at 39°(Table 1.). Thus, under conditions which could not have selected for loss of the ts genes, their presence did not inhibit the growth of the hybrid cells at 39°, when in the presence of a ts^+ allele.

These experiments proved conclusively the recessive nature of our ts mutations. Thus, one of the hypotheses mentioned before to explain the expression of recessive mutations in a diploid cell must be considered. At present we favor the second interpretation, i.e., that our mutations occur in genes, that because of a preexisting or subsequent event, became haploid in the mutant cells. This is supported by the fact that most of our mutants seem to have a chromosome number which is lower than that of the parental cells (unpublished observations), and by the fact that we have been able, in some cases, to rule out the possibility that our mutations are localized on the X-chromosome. This last type of evidence derives from experiments with the ts HGPRT⁻ BHK mutants. The genes for the HGPRT enzyme are located on the X-chromosome. Therefore we carried out linkage experiments with these cells to ascertain whether their ts mutation were linked to the HGPRT marker and therefore located on the X-chromosome. We have tested 5 mutants and in all cases no linkage was found between the HGPRT marker and the ts marker (Meiss and Basilico, in preparation).

Having proven the recessiveness of our ts mutations, we attempted complementation analysis of the mutants. Such complementation studies can tell whether mutants independently isolated and of unknown phenotype represent mutations in the same or in different genes. We have used the technique of somatic cell hybridization as described before. As of now the mutants tested fall into 11 complementation groups, which are seen in Table 2. As it can be seen, most mutants obtained are unique, that is, most complementation groups contain only one mutant.

Table 2. Complementation Groups of ts BHK Mutants

1	2	3	4	5	6	7	8	9	10	11
BCH	AF8	422E	AF6	T22	T15	T23	TD	T14	T2	N5
BCL						T60				B2-73
BCB										B2-63

Characterization of the ts Mutants.
  We will describe here the properties of two of our ts mutants, ts 422E and ts AF8, which have both been extensively characterized.
  Ts 422E. The ts 422E mutant cannot grow at 39° because of a block in the production of 28S ribosomal RNA or 60S ribosomal subunits (Figure 1.). This defect is evident almost immediately after shift to the non-permissive temperature, but does not reach its maximum until about 16 hours after shift up. Cell growth also stops at about the same time, so that on average the cells at 39° perform one division (10).
  DNA synthesis or overall protein synthesis are not grossly impaired even 4 days after shift to 39°. Production of 18S rRNA, or of small ribosomal subunits is normal (10). 40S subunits produced at 39° are fully functional, as they are capable of exchanging with old 40S subunits in 80S ribosomes, and are found on polysomes. The proportion of newly formed 40S subunits that, following a two hours label, are found "free" in the cytoplasm is only about double that of the wt cells (Table 3.). This increase is hardly surprising, since concomitant production of the large ribosomal subunits does not take place.
  The mechanism responsible for the lack of production of 28S rRNA at 39° resides in a defective processing of the rRNA precursors. This is demonstrated by a) the lack of appearance of 28S RNA in the nucleolus or the nucleoplasm of ts 422E cells

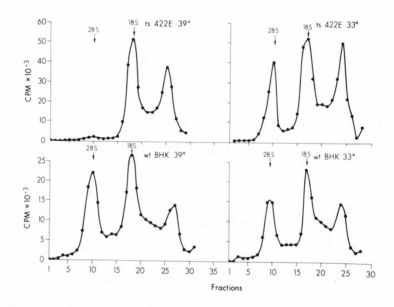

FIG. 1.   Production of cytoplasmic rRNA in ts
422E and wt BHK cells at 33° and 39°.   Cells were
plated at 33° and 39°.   After 2 days they were
labeled for 2 hr. at 33° and 1.5 hr. at 39° with
(^{3}H)uridine (1 μCi, 0.2 μg/ml).   The cytoplasm
from 1x10^7 cells was analyzed on 15-30% SDS-sucrose
gradients.   Sedimentation is from right to left.
The arrows indicate the position in the gradients
of the A 260 nm peaks of 28S and 18S rRNA, obtained
from collection of the gradients through an auto-
matically recording Gilford spectrophotometer.
From Ref. (10).

TABLE 3. Fate of Newly Made Ribosomal Subunits in the Cytoplasm of ts 422E Cells at 39° *

Subunits	Cells	% of total radioactivity incorporated in subunits		
		in 80S ribosomes on polysomes	on monosomes	in "free" subunits
60S	wt BHK	32.5	31.5	36
	ts 422E	-	-	-
40S	wt BHK	24.0	57.0	29
	ts 422E	13.5	27.5	59

*ts 422E and wt BHK cells were shifted to 39°, and 2 days later labeled with 3H-uridine (1μCi, 0.2μg/ml) for 2 hours. The cytoplasmic fraction was prepared and divided into 2 parts. One half was treated with RNase (2μg/ml, 10' at 0°C) and then analyzed on 15-30% sucrose gradients in RSB buffer (20,000 rpm for 18 hrs. at 5°) to obtain the distribution of radioactivity in 80S ribosomes and free 60S and 40S subunits. The other half was sedimented on 15-30% sucrose gradients in RSB (24,000 rpm for 3 hrs. at 5°C). The polysomes containing fractions were pooled, treated with 1% SDS, and analyzed on 15-30% SDS-sucrose gradients to obtain the distribution of radioactivity in the two ribosomal subunits on polysomes. The percent of newly made subunits found on 80S monosomes was obtained by subtracting the polysomes value from the value obtained for total 80S ribosomes.

at 39°; b) no evidence of production and degradation; c) an accumulation of the 32S RNA precursor in the nucleolus of ts 422E cells at 39° (10).

The accumulation of 32S RNA reaches levels about 10 times higher than those of wt cells (10). However, it fails to account for the whole amount of 28S RNA which is not produced. When ts 422E cells are labeled at 39°, the cells fractionated into nucleolar and cytoplasmic fractions, and the RNA analyzed on sucrose gradients or polyacrylamide gels, it can be observed that: 1) Synthesis of the large rRNA precursor molecule (45S) appears to be normal. 2) With pulses up to 30' processing appears to be normal, except for the appearance of an intermediate precursor, tentatively identified as 35S, which is almost undetectable in wt cells. 3) Longer pulses, or chases of radioactivity, reveal a disappearance of the 35S precursor and some accumulation of 32S RNA. The amount of 32S RNA increases slowly with time. Most of it, however, is degraded, as it is not processed to 28S RNA. 4) When cells are labeled for longer periods (3-5 hours) and the radioactivity chased, it can be seen that of the accumulated 32S, some is slowly processed to 28S RNA, but the largest part is also eventually degraded (Figure 2.).

The simplest interpretation of these results seems to be the following: ts 422 has an aberrant processing of the 32S rRNA precursor. This leads to degradation or accumulation. This could be due to a defect in a processing enzyme, or to the formation of an aberrant 32S RNA containing preribosomal particle. The second hypothesis is supported by the following experiment: When ts 422 cells are labeled at 39° for 3 hours, and the radioactivity is chased for 6-20 hours, it can be seen that very little 28S rRNA is made, and most of the 32S RNA is degraded, although as late as 20 hours after the pulse a proportion of it can still be found in the nucleolus. This pattern is essentially unchanged when the chase of radioactivity is performed at 33°. Even 20 hours after shift down, when the processing of the newly synthesized rRNA has returned to normal (10), some of the 32S RNA synthesized at 39° is still found in the nucleolus (Figure 2.).

These data strongly suggest that the particle containing 32S RNA is defective, so that it cannot

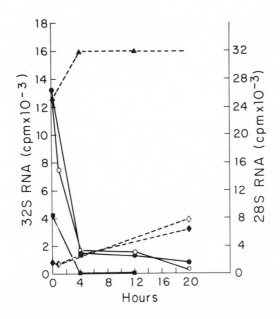

FIG. 2. Conversion of the 32S rRNA into mature 28S rRNA in ts 422E cells at 39° and 33°. ts 422E and wt BHK cells were shifted to 39°. 30 hrs. later they were labeled for 3 hrs. with (^{3}H)methionine (10 μCi, 5 μg/ml). At the end of the pulse (0 time), part of the cells were fractionated into nucleolar and cytoplasmic fraction, the RNA extracted and analyzed on polyacrylamide gels as described (10). The remaining portion of the cells were washed and incubated in medium containing an excess of cold methionine (6 mg/ml) either at 39° or 33°. At the times indicated they were fractionated and the RNA analyzed as before. Radioactivity (cpm) in cytoplasmic 28S RNA: ▲-▲ wt BHK, chase at 39°;◆-◆ ts 422E, chase at 39°;◇-◇ ts 422E, chase at 33°. Radioactivity (cpm) in nucleolar 32S RNA:■-■ wt BHK, chase at 39°; ●-● ts 422E, chase at 39°; o-o ts 422E, chase at 33°. All cpm are normalized for the same amount of radioactivity in 18S rRNA.

be properly utilized even when the cells are shifted to the non-permissive temperature. A likely
reason for this behavior would be the presence of a
ts ribosomal protein. Attempts to identify this
protein are under way.

Aside from the importance for the study of the
biosynthetic processes involved in ribosome production, the ts 422E mutant should also be useful for
studies on the relationship between cell proliferation and rRNA synthesis and protein synthesis. In
fact, a striking property of this mutant is that
although cell division stops about 12 hours after
shift up, protein synthesis (determined as incorporation of radioactive aminoacids into acid insoluble material) continues for long periods of time
(10). Thus, cessation of growth does not result
from an arrest of protein synthesis. It is therefore possible that production of ribosomes is
tightly correlated with the orderly progression of
the cell through the mitotic cycle, or mitosis it-
self. Alternatively, it could be thought that a
normal amount of ribosomes per cell is critically
related to cell division, influencing the synthesis
of some specific proteins. Experiments to distinguish between these hypotheses are under way.

Ts AF8. The result of specific ts defects on
the progress of the cell through the mitotic cycle
has been also studied with another ts BHK mutant,
ts AF8.

Ts AF8 grows normally at 33°, but upon shift to
39° growth stops after about 12-24 hours and the
cells perform on average less than one division.
The cells remain viable at 39° for long periods of
time (Figure 6.). DNA synthesis is reduced gradu-
ally upon shift up, primarily because of a decrease
in the frequency of DNA synthesizing cells, rather
than a decrease in the rate of DNA synthesis/cell
(11).

We measured the rate of entry into S of ts AF8
at 39°. In Figure 3., it can be seen that at 39°
the rate of entry into S of ts AF8 is, extrapolat-
ing from the initial portion of the curve, at least
30 times slower than that of wt BHK.

These data showed that ts AF8 cells at 39° are
arrested in their progression through the cell
cycle. We were able to show that ts AF8 cells become arrested in the G1 phase of the cell cycle by

86

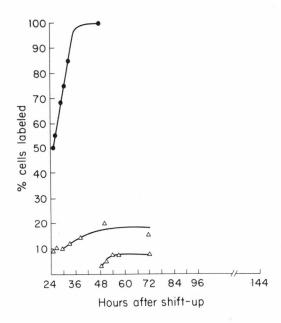

FIG. 3. The rate of entry of ts AF8 cells and wt BHK cells into the S phase of the cell cycle. Cells were plated at 33° and allowed to grow for at least 24 hrs. before they received fresh medium and were shifted to 39.5°. Cells were incubated with ^{3}H dT (3 μCi, 0.5 μg/ml), which was added either 24 or 48 hrs. after shift up. At the times indicated the cells were fixed for radioautography. For each point at least 500 cells were counted. ●-● BHK; Δ-Δ ts AF8.

determining the distribution of DNA content in ts AF8 populations at different temperatures. Figure 4. shows that at 39° ts AF8 has an essentially unimodal distribution, with a DNA content equal to that of G1 cells at 33°.

Following these conclusions, we determined whether ts AF8 had a defect which led to the arrest of cells anywhere in G1, or whether the cycle arrest point of the mutation (i.e., the point in the cell cycle where the mutation led to an interruption of the normal growth progression) was localized in a specific part of the G1 phase. We used various methods of inducing synchrony coupled with shifts to the non-permissive temperature (11).

The results can be summarized as follows (Table 4.): 1) Cells arrested at the G1/S boundary by a hydroxyurea window technique (11), and shifted to 39° at the time of removal of hydroxyurea, enter S normally. Therefore the block must be prior to the G1/S boundary. 2) Cells synchronized by isoleucine deprivation, which is believed to block cells in G1 (12), also enter S normally if shifted to 39° at the time of addition of isoleucine. 3) Cells synchronized by serum starvation, which blocks cells in early G1 (G0), do not enter S if shifted to 39° at the time of serum addition. However, if these cells are allowed to progress through G1 and are shifted to 39° shortly before entering S at 33°, then they enter and complete S normally (Table 4.).

These data allow us to place the cycle arrest point of the ts AF8 mutation on a cell cycle map between the serum block proximal to, and the isoleucine block distal to the cycle arrest point of the mutation (Figure 5.).

It was considered of interest to investigate whether the synthesis of extrachromosomal DNA was affected by the ts AF8 mutation. We determined the rate of synthesis of mitochondrial DNA in ts AF8 and BHK cells at 39°. As late as 48 hours after shift up, mitochondrial DNA synthesis in ts AF8 proceeds at wt rates (11). We have also carried out infection of ts AF8 cells with vaccinia virus. The results show that the replication of vaccinia virus DNA, which is known to take place in the cell cytoplasm, was unaffected by temperature (Table 5.). Thus the ts AF8 mutation does not affect synthesis of extrachromosomal DNA.

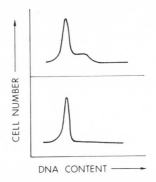

FIG. 4. Distribution of DNA content in ts AF8 populations at 39° and 33°. ts AF8 cells growing at 33° (top) or 48 hrs. after shift to 39° (bottom) were fixed with methanol and prepared for modified fluorescent Feulgen measurement of cellular DNA, performed by flow microfluorometry (15), which was kindly done by Dr. X. Yataganas (Sloan-Kettering Institute, New York City). The figure shows that the ts AF8 population at 39° has a unimodal distribution of DNA content equal to that of G1 cells at 33°.

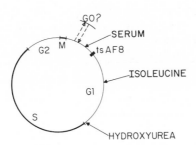

FIG. 5. Cell cycle map, indicating the presumptive position of the ts AF8 arrest point in the G1 phase of the BHK cell cycle. Hydroxyurea: block induced with a hydroxyurea window technique (11). Serum: block induced by serum starvation. Isoleucine: block induced by isoleucine deprivation. The precise location of these blocks is arbitrary. Only their order is meaningful.

TABLE 4. Effect of Shift Up on
Synchronized ts AF8 Cells

Synchrony induced at 33° with	Time[†] of Shift to 39°	Temperature	% DNA synthesizing cells[#]
hydroxyurea*	-	33°	62
	0	39°	58
serum starvation**	-	33°	67
	0	39°	8
	26 hrs.	39°	67
ileu deprivation***	-	33°	80
	0	39°	75

* A window technique which accumulates cells at the G1/S boundary (11).
** 0.5% serum for 52 hrs.
*** Incubation in medium without isoleucine for 48 hrs.
† After removal of the synchronizing block.
# Maximum frequency of DNA synthesizing cells, obtained by pulsing the cells with ^{3}H-thymidine for 1 hr. at various times after removal of the synchronizing block. Values were determined by autoradiography.

TABLE 5. Synthesis of Vaccinia Viral
DNA in ts AF8 Cells*
Radioactivity (cpm/10^6 cells) Incorporated
into Cytoplasmic DNA

Temperature	Hours after infection	wt BHK	ts AF8
39°	2	110	470
	6	2110	5360
33°	3	210	520
	6	1400	3300

* ts AF8 and wt BHK cells, growing at 33° or 36 hrs. after shift to 39° were infected with Vacinnia virus (strain WR) at 150 particles/cell. At the end of the adsorption period the cells were incubated in ^{14}C-thymidine (0.1μCi,0.4μg/ml) At the times indicated the cytoplasmic fraction was prepared, TCA precipitated, filtered on Millipore filters, and the radioactivity counted.

With this mutant, we have performed experiments of transformation or infection with Polyoma virus, an oncogenic virus capable of inducing cellular DNA synthesis in resting cells and of transforming BHK cells. Polyoma transformation does not cause phenotypic reversion, as Polyoma transformed AF8 (Py ts AF8) are still incapable of growth at 39°. In addition, infection with Polyoma virus (200-500 PFU/cell) of ts AF8 at 39° does not induce DNA synthesis or abortive transformation. However, there are some differences between the behavior of Py ts AF8 and their untransformed counterpart. When ts AF8 cells are shifted to 39°, their c-AMP concentration increases, and ribosomal RNA synthesis decreases to about 1/3 the wild-type rate (11); protein synthesis continues, but the cells do not accumulate proteins. All these facts are consistent with a balanced arrest of growth in the G1 phase of the cell cycle. Py ts AF8, on the other hand, do not show increased c-AMP levels, and the cell protein content increases. Most important, while ts AF8 cells remain viable at 39° for many days, Py ts AF8 lose viability at a very fast rate (Figure 6.). This is not due to induction of virus production.

This phenomenon is of interest since it suggests that transformed cells respond to a specific biochemical block in a different way from normal cells. The different response does not consist in overcoming the effect of the mutation, as the entry into S of Py ts AF8 cells at 39° is still impaired. It rather seems that these cells have lost the capacity to respond to a G1 block with a balanced growth arrest. Death may thus result from unbalanced growth. If the ts AF8 cells at 39° reached a G0 state, it could be thought that transformed cells have lost the capability of entering this subset of the cell cycle. The hypothesis that transformed cells do not respond to physiological stimuli in a normal fashion has been advanced before. Our results suggest that transformed cells might lose the ability of arresting their growth in a balanced way (e.g., entering G0), irrespective of the type of stimulus which would cause growth arrest.

This matter needs to be investigated further. The G0 concept is not very well defined, and it is

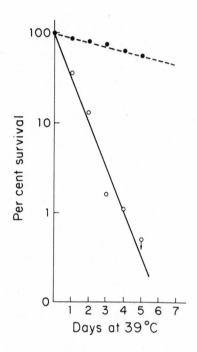

FIG. 6. Survival of ts AF8 cells and Polyoma transformed ts AF8 at 39°. Cells were plated at 33°. After 6 hrs. they were shifted to 39°. At the times indicated some cultures were shifted to 33° and allowed to grow for an additional 14 days. Survival was determined on the basis of the ability of the cells to form colonies at 33°. ●-● ts AF8; o-o Py ts AF8 (clone #8).

not yet clear whether it really represents a different phase from G1. Further studies should help elucidate these differences, and the ts AF8 mutant should be useful for this purpose.

Conclusions

The main thrust of the results we presented is to demonstrate that ts mutants of mammalian cells can be obtained and that they behave in a reproducible manner, consistent with the theories about the origin of these mutations which have been derived from studies with other organisms. Their reversion frequency can be increased by mutagens, the mutations are generally recessive and chromosomal in nature. All these data support the idea that they derive from alterations in specific proteins, which are functional at low but not at high temperature.

Complementation analysis can be performed, and the usefulness of these mutants for studies of somatic cell genetics is beginning to be demonstrated. The main obstacle to the use of these mutants for studies of cell physiology and growth regulation is that biochemical characterization is rather difficult. However, the mutants can be characterized in a satisfactory manner as demonstrated by the examples described in this paper.

The study of these mutants shows that specific genetic defects can inhibit the progression of the cell through the cell cycle and influence its growth regulation. The ts 422E mutant, which has a temperature-sensitive defect in the production of ribosomal RNA, should be useful for studies on the relationship between ribosomes production, or concentration, and cell growth. The other mutant, ts AF8, belongs probably to a category of mutants which are going to be fairly common. Two ts mutants of Chinese hamster cells which probably become arrested in G1 at the non-permissive temperature have been described (13,14), and we are studying another ts BHK mutant, ts N5, which appears also to be blocked in G1. Ts AF8, however, complements with ts N5 and with the ts mutant of Scheffler and Buttin (13), as we have determined by cell hybridization.

The availability of several ts mutations, all affecting the cell progression through G1, should be of great advantage for the study of this poorly understood phase of the cell cycle. Since it is probably in G1 that most of the controls regulating cell growth are effective, this should help elucidate the control of proliferation in animal cells, and their alterations in cancer cells.

Another type of study in which these cells should be useful is the investigation of the host cell functions required for viral multiplication, or neoplastic transformation. We have initiated, in collaboration with Dr. H. Raskas (Washington University, St. Louis), experiments aimed at determining the capacity of several ts BHK mutants to support the multiplication of Adeno virus 2 at the non-permissive temperature. Preliminary results do indeed show that Adeno virus multiplication does not take place in some of the mutants at 39°. The continuation of these studies should help identify the host cell functions which are required for the multiplication of this oncogenic virus. With regard to viral transformation, on the other hand, the use of ts cell mutants should allow determining the requirements for specific cellular functions for a number of basic events in transformation, such as the integration of the viral DNA into the cellular genome, etc. Such an approach should provide much more solid data on these requirements than those obtained with the use of metabolic inhibitors, which do not discriminate between viral and cellular biosynthetic processes.

In conclusion, the data presented in this paper suggest that an approach to the understanding of cellular physiology with the use of genetic tools, such as our ts mutants, should contribute valid knowledge on a number of problems, including growth regulation and cancer.

This approach is obviously a long term one, as the creation of a genetic system of somatic animal cells will probably require a long time. Nevertheless, we feel that without such system, the fine mechanisms which regulate cell growth and division will be difficult to understand.

Acknowledgements
This investigation was supported by PHS Contract NO1 CP33279 from the National Cancer Institute. C.B. is a scholar of the Leukemia Society.

References

1. Epstein, R.H., Bolle, A., Steinberg, E.M., Kellenberger, E., Boy de la Tour, E., Chevelley, R., Edgar, R.S., Susman, M., Denhardt, G.H., and Lielausis, I. Cold Spring Harbor Symp. Quant. Biol. 28, 375 (1963).
2. Edgar, R.S., and Lielausis, I. Genetics 49, 649 (1964).
3. Renger, H.C.,and Basilico, C. Proc. Nat. Acad. Sci. USA 69, 109 (1972).
4. Renger, H.C., and Basilico, C. J. Virol. 11, 702 (1973).
5. Stoker, M., and Macpherson, I. Nature 203, 1355 (1964).
6. Meiss, H.K., and Basilico, C. Nature New Biol. 239, 66 (1972).
7. Basilico, C., and Meiss, H.K. Methods in Cell Biol. 8 (1974), in press.
8. Hartwell, L.H. J. Bacteriol. 93, 1662 (1967).
9. Suzuki, D.T., and Procunier, D. Proc. Nat. Acad. Sci. USA 62, 369 (1969).
10. Toniolo, D., Meiss, H.K., and Basilico, C. Proc. Nat. Acad. Sci. USA 70, 1273 (1973).
11. Burstin, S.J., Meiss, H.K., and Basilico, C. (to be published in J. Cell.Physiol.)
12. Ley, K.D., and Tobey, R.A. J. Cell Biol. 47, 453 (1970).
13. Scheffler, I., and Buttin, R. J. Cell. Physiol. 81, 199 (1973).
14. Roscoe, D.M., Robinson, H., and Carbonell, A.W. J. Cell. Physiol. 82, 333 (1973).
15. Kraemer, P.M., Deaven, L.L., Crissman, M., and Van Dilla, M.A. in "Advances in Cell and Molecular Biology" ed. E.J. Dupraw, 2, 1972.

FLUID LIPID REGIONS IN NORMAL AND ROUS SARCOMA VIRUS TRANSFORMED CHICK EMBRYO FIBROBLASTS

Betty Jean Gaffney[a], P.E. Branton[b], G.G. Wickus[b], and Carlos B. Hirschberg[b]

Submitted by Phillips Robbins

[a]Stauffer Laboratory for Physical Chemistry, Stanford University, Stanford, California; Present Address: Department of Chemistry, The Johns Hopkins University, Baltimore, Maryland

[b]Department of Biology, Massachusetts Institute of Technology, Cambridge, Massachusetts

Comparison of the lipid fluidity in normal and transformed cells is of interest for several reasons. One reason is that numerous investigations have shown a relationship between transformation and enhanced agglutinability by lectins (1). Both normal and transformed 3T3 cells have a dispersed arrangement of cell surface lectin receptors at $4°C$. At $37°C$, the receptors migrate to form clusters on transformed cells but remain dispersed on normal cells (2,3,4). Because of this temperature dependence, it has been suggested that the fluidity, or viscosity, of membrane lipids may be a factor controlling the differential redistribution of lectin receptors and the differential agglutinability of normal and transformed cells. A second reason for comparing lipid fluidity in normal and transformed cells is that even if lipid fluidity does not <u>control</u> the mobility of cell surface receptors, a change in the physical state of some part of the transformed cell membrane could be translated into a detectable change in the overall fluidity of the

membrane lipids. Recent investigations have shown
that the interaction of red blood cell membranes
with very low levels of hormones (5) and other
small molecules (6) produces a measurable change
in the "fluidity" of the membrane lipids. We
report here our investigation of the membrane
fluidity of normal and SR-Rous Sarcoma Virus
transformed chick embryo fibroblasts (CEF). The
term "membrane fluidity" includes a wide range of
molecular motions. In the present study, we used
fatty acid spin labels which are particularly
sensitive to the flexibility of the hydrocarbon
chains of membrane lipids(7,8).

We employed spin label fatty acids, I (m,n),
as probes of the plasma membrane lipids because
1) these labels are quite sensitive to changes in

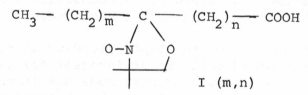

$$CH_3 — (CH_2)_{\overline{m}} \; C \; — (CH_2)_{\overline{n}} — COOH$$

$$O-N \qquad O$$

I (m,n)

the mobility of membrane lipids (7,8), 2) the
paramagnetic resonance spectra of these labels are
well understood (8,9) and 3) measurement of lipid
fluidity may be made in a matter of minutes with
relatively low concentrations of label. RSV
transformed chick cells are particularly attractive
because a mutant of RSV (TS-68) is available
which is temperature sensitive for transformation
but not growth. Infected cells grown at $41°$ have
the normal phenotype whereas those grown at $36°$
display the transformed phenotype (10). The
CEF-RSV cells were also attractive because the
changes in lipid composition accompanying transformation are minimal (11), implying a more subtle
basis for drastic changes in membrane structure.
The selective agglutination of transformed CEF-
RSV cells by lectins has been observed (12, 13, 14).
The possibility that the fluidity of membrane lipids might control the mobility of membrane

proteins and cell surface receptors is suggested
by several observations besides the agglutination
experiments mentioned above. The measured rates
of rotation and lateral motion of rhodopsin in rod
disc membranes are consistent with estimates of
the viscosity of fluid phospholipid bilayers (15,
16); changes in the activity of certain bacterial
membrane proteins may be evaluated in terms of
phase diagrams describing phase separations of the
membrane lipids (17); changes in the rate of
mixing of cell surface antigens of mouse-human
heterokaryons as a function of temperature are consistent with lipid phase separations and with
changes in lipid fluidity (18); and anti-
immunoglobulin induced patch formation on lymphocytes is temperature dependent and occurs in the
presence of azide (19,20). The factors which
produce demonstrable changes in the flexibility of
fatty acid spin labels in lipid bilayers include
temperature (21), fatty acid composition (22),
phospholipid to cholesterol ratio (23), the
presence of membrane proteins (24), and as mentioned above, the reversible binding of certain
hormones (5) and small molecules (6) to red
blood cell membranes.

Methods

CEF were prepared from 11 day old embryos
(COFAL-negative, C/O; SPAFAS, Norwich, Conn.) as
described by Rein and Rubin (25). These primary
cultures, seeded at a density of $1X10^7$ cells per
100mm plastic petri plate (Falcon), were infected
three hours after preparation with $5X10^6$ FFU of
SR-RSV(A) or the thermosensitive derivative TS-68
(10). The cultures were grown for 2 1/2 days in
Medium 199 supplemented with 2% tryptose phosphate
broth, 1% calf serum (GIBCO) and 1% heat-
inactivated chicken serum (GIBCO). The TS-68
infected primary cells were cultured at 41^c. The
uninfected and SR-RSV(A) infected primary cells
were grown at 39^c.

After 2 1/2 days, both infected and uninfected cultures were trypsinized and transferred. Each cell type was plated at a density of $5X10^6$ cells per 100mm petri dish in Medium 199 supplemented with 10% tryptose phosphate broth, 4% calf serum, and 1% heat-inactivated chicken serum. Secondary cell cultures of each type were maintained at both 36^c and 41° . Full morphological transformation occurred in the SR-RSV(A) infected cultures and the TS-68 infected cultures maintained at 36^c 48 hours after transfer from the primary cultures. TS-68 infected cultures maintained at 41^c retained a normal morphology. Cells were harvested and spin label was added at the onset of full morphological transformation as well as 24 and 48 hours there-after. Throughout this study the medium was changes every 12-18 hours.

Fatty acid spin labels, I (m,n) were synthe-sized as previously described for (m,n) = (10,3); (7,6); (5,10) and (1,14). (8)   Before intro-duction of the labels into cells, a cell monolayer was washed on the dish three times with Solution A (calcium, magnesium-free PBS)(27) at 37^c . After the last wash was removed by decanting, the mono-layer was gently pealed to one side of the culture dish with a rubber policeman. The fatty acid spin label in ethanol solution (2 μl of ~$2X10^{-2}$ M solution) was allowed to evaporate on the bare side of the dish.   The cells took up the spin label rapidly as they were gently pushed over the residue of spin label for about 30 seconds.

Paramagnetic resonance signals of labeled cells were recorded on a Varian E-9 spectrometer equipped with a variable temperature controller. The samples consisted of about $6X10^6$ to $1X10^7$ cells contained in 50 μl glass disposable pipettes (Corning) sealed at one end. Spectra were usually recorded in within 15 minutes after introduction of the spin label. When the spectra of several samples were recorded repeatedly over one hour, however, no changes in the order parameter of the

100

sample were observed. During the one hour period, there was considerable reduction in amplitude of the signal, presumably due to chemical reduction of the paramagnetic nitroxide group by components of the cell cytoplasm. Maximum spectral sensitivity was obtained by recording the inner spectral extrema ($T_\perp'$ (approx.)) at a 40 gauss scan width. Modulation amplitudes were one gauss or less and the microwave power was kept at or below 10mW to avoid sample heating and signal saturation. The temperature was monitored by using a copper-constantin thermocouple placed just above the resonance area of the cavity. No gradient of temperature was observed when the thermocouple was lowered one cm. into the cavity to the position of the sample. For a single set of measurements comparing normal and transformed cells, the settings on the temperature controller and the position of the sample holder were not varied.

For most comparisons of normal and transformed cells, only the values of $T_\perp'$ (approx.) were recorded. This parameter varies linearly with order parameter, S, for lipid spin labels in lipid bilayers (26). For the data shown in Fig. 2, both $T_\parallel'$ and $T_\perp'$ (approx.) (half the separation of the outer and inner spectral extrema, respectively) were recorded and order parameters, S_n, were calculated using the following formula (26):

$$S_n = \frac{T_\parallel' - (T_\perp' \quad (approx.) + C)}{T_\parallel' + 2(T_\perp' \quad (approx.) + C)} \times 1.723$$

Here $T_\perp'$ (approx.) is the value obtained directly from the experimental spectrum. This parameter deviates slightly from the value of $T_\perp'$ obtained from computer calculation of spectra to fit the experimental data. The correction term which must be added to $T_\perp'$ (approx.) increases as the order parameter decreases and is given by:

$$C = 1.4 \text{ gauss} - 0.053 (T_\parallel' - T_\perp' (approx.)) \text{ gauss}$$

Order parameters as low as 0.2 may be calculated with reasonable accuracy by the above method (26).

Plasma membranes were isolated from normal and RSV-infected chick embryo fibroblasts using a modification of the method of Brunette and Till (27). In brief, the cells, growing in 100mm plastic dishes, were washed three times with about 10 ml of warm PBS (phosphate buffered saline) (28). They were scraped into PBS with a rubber policeman and pelleted by low-speed centrifugation. The cells were then resuspended in 10^{-3}M tris (pH 7.5) containing 5×10^{-4} M $ZnCl_2$. The cell concentration used was approximately 10^8 cells per 10 ml of tris-$ZnCl_2$ buffer. The suspension was incubated at room temperature for 15 minutes and then in ice for an additional 15 minutes. The cells were ruptured in a Dounce homogenizer with a type B pestle (Kontes, Vineland, N.J.) until about 90% of the cells had been broken. The homogenate was sedimented for 15 minutes at 1400 rpm in the 269 head of an International PR2 centrifuge. The resulting pellet was then fractionated in an aqueous polyehtylene glycol-dextran 2-phase system as described by Brunette and Till (27), except that no $ZnCl_2$ was included. Plasma membrane, present at the interface of the two phases, was collected and washed several times with water. Such membranes typically showed a 6 to 10-fold increase in the specific activity of 5'-nucleotidase (Branton, unpublished). In addition they contained very low levels of DNA and RNA, as measured by ^{3}H-thymidine and ^{3}H-uridine labelling, and of endoplasmic reticulum, as measured by DPNH-diaphorase activity (Branton, unpublished).

For extraction and analysis of lipids, growth medium was removed from cells and each plate was rinsed three times with 5 ml of cold 0.01 M Tris hydrochloride, pH 7.5 (0.15M NaCl). This solution (3 ml) was then added to each plate and the cells were scraped off the plate with a rubber policeman and transferred to a glass centrifuge tube. The

plates were rinsed two more times and the suspension was centrifuged for 5 min. at 600 xg and the
supernatant was discarded. The cell pellet was
resuspended in 3 ml of this solution and
recentrifuged. The cells were then suspended in
1 ml of methanol and the suspension was dried
under a stream of nitrogen at warm temperature.
Lipids were extracted with redistilled chloroform-
methanol 2:1 and finally with chloroform-methanol
1:2. The dried samples were dissolved in 2 ml of
2.5% sulfuric acid in redistilled methanol and
refluxed at 80° for 6 hours and the fatty acid
methyl esters were extracted into hexane as described by Smith et al. (29). The fatty acid methyl
esters were examined by GLC as previously described
(30). Plasma membrane lipids were determined as
described for whole cells.

Results

Phase separations of lipids have been observed
in biological membranes with simple and well
defined lipid composition (17). In cell membranes
with more complex lipid composition, there are
undoubtedly separations of fluid and solid domains
of lipids although the details of the phase diagrams for these separations are probably quite
complex. When a fatty acid spin label is present
at a low concentration in bilayers which have both
solid and fluid domains the label may partition
preferentially into the fluid domains of the lipid
bilayer. That this indeed occurs is demonstrated
by Figure 1 for mixtures of dipalmitoyl lecithin
(DPL) and dioleoyl lecithin (DOL). Here, the lipid
fluidity was measured as a function of temperature.
The fluidity is expressed in terms of $(2\ T_L')^{-1}$, a
spectral parameter related to the order parameter,
S_n, of the fatty acid spin label. This order
parameter is a measure of the amplitude of motion
of fatty acid chains and has a value of 1.0 in a
solid and approaches zero when the label is in a

very fluid environment. In pure dipalmitoyl
lecithin, the motion of the fatty acid label I
(5,10) responds to the characteristic phase
transition of DPL. The response is an abrupt
decrease in order parameter below $\sim 40^{c}$ C (Figure 1,
curve a). When the same label is present in bilayers of DOL the order parameter also decreases
with temperature as shown in Figure 1 d. However
because the phase transition of DOL occurs below
the temperatures shown, there is no abrupt change
in order parameters in curve d. In a 3:1 or 1:1
mixture of DPL and DOL (curves b and c), the
temperature dependence of the order parameter
shows that the motion of the spin label responds
slightly to the onset of the separation of a solid
phase which must be highly enriched in the DPL
component. At temperatures below the onset of the
phase separation for curves b and c, the label
clearly resides in a phase which is only slightly
less fluid than pure DOL. This phase must have a
lower ratio of DPL to DOL than the ratio for the
total sample (the solid phase must therefore have
an elevated DPL to DOL ratio). Although complete
understanding of the data of Figure 1 requires a
phase diagram (21) for the DOL-DPL lipid system,
the data do indicate that the fatty acid spin
label partitions preferentially into the fluid
regions of this bilayer system in the temperature
range shown. In addition, phospholipid molecules
exhibit rapid lateral diffusion in fluid lipid
bilayers (31) and in biological membranes com-
posed of both lipids and proteins (32). The
fatty acid spin labels certainly must have high
lateral mobility as well and will thus sample an
average environment. Therefore it seems reasonable
to conclude that the fatty acid spin labels in
both cell membranes and in synthetic phospholipid
bilayers reflect the <u>average properties of the
fluid lipid regions of the membranes</u>.

Before considering the possibility that viral-
transformation may involve a change in average

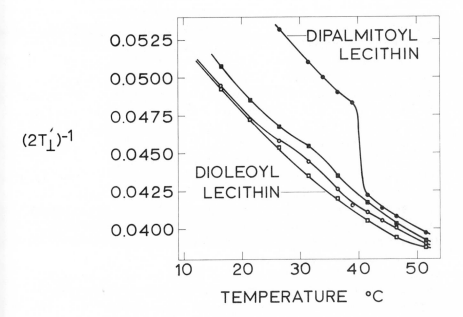

Figure 1. A plot of $(2T_{\perp}')^{-1}$ as a function of temperature for 0.5 moles of lecithin spin label I (5,10) in a 20% suspension of lipid in water. Curves from upper to lower are for (a) dipalmitoyl lecithin (synthetic L-α-dipalmitoyl lecithin from Calbiochem), (b) 3:1 mixture (mole/mole) of dipalmitoyl lecithin-dioleoyl lecithin (L-α-dioleoyl lecithin from Supelco, Inc.) (c) 1:1 mixture of dipalmitoyl lecithin-dioleoyl lecithin, (d) dioleoyl lecithin.

properties of fluid regions of the cell membranes
of CEF we will compare the membrane fluidity, as
measured by the fatty acid spin labels, in
different cell types. Figure 2 shows a plot of
the order parameter of fatty acid spin labels as
a function of increasing number of methylenes,
($-CH_2$), between the carboxyl group and the
position of the spin label group on the hydrocarbon
chain. Plots are given for human red blood cells
(HRBC) and CEF. Clearly, the fluid lipid
region of the HRBC membrane is much more rigid
than of the chick cell membranes at the
temperature of these measurements ($38°C$). The large
differences in fluidity of HRBC and CEF may
reflect differences in cholesterol: phospholipid
or total lipid: protein ratios, or more subtle
differences in membrane properties.

Repeated measurements of lipid fluidity
were made using label I (7,6) in uninfected CEF,
S R-RSV(A) transformed CEF and TS-68 infected CEF.
Cells of each type were grown both at $36°$ and $41°$.
The permissive temperature for expression of the
transformed phenotype in TS-68 infected CEF is $36°$.
The TS-68 infected cells grown at $41°$ appeared
normal for all experiments reported. The summary
of data obtained from these measurements is
presented in Table I. The average of several
measurements for each cell type is given for three
separate sets of experiments. Each experiment
utilized cells which were prepared at different
dates. Although the data of Table I indicate
that the TS-68 infected cells at both $36°$ and $41°$
may be slightly more rigid than the corresponding
normal cells, there is no difference in fluidity
which can clearly be associated with transformation
when all of the data is considered.

Plasma membranes were also prepared from
normal and transformed cells by the procedure of
Brunette and Till (26). No consistent
differences were found in the fluid lipid regions
of normal and transformed plasma membranes either.

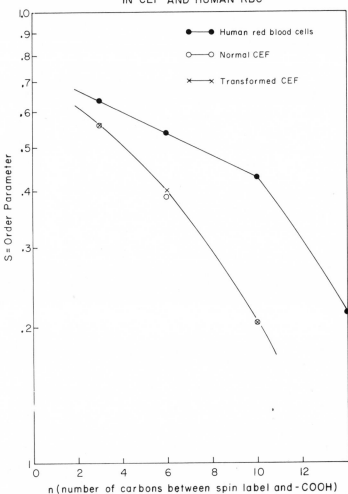

ORDER PARAMETER vs. n FOR FATTY ACID SPIN LABELS
IN CEF AND HUMAN RBC

Human red blood cells
Normal CEF
Transformed CEF

S = Order Parameter

n (number of carbons between spin label and - COOH)

Figure 2. The order parameter, S_n, as a function of the number, n, of methylenes $(-CH_2-)_n$ separating the carboxyl and spin label groups on the spin label fatty acids I (m,n). Spectra were recorded at 38°C for intact human red blood cells and normal and RSV-transformed chick embryo fibroblasts. The data for I(7,6) in normal and transformed cells were selected to show the magnitude of variation in measurements reported in Table I.

Table I. Measurements of Lipid Fluidity[a] in
Normal and Transformed CEF

	growth temperature	CEF(b)	CEF-RSV(b)	CEF-RSV-TS-68(b)	temperature of spectral measurement
Experiments A	36°	10.55(3)	10.55(3)	10.38(3)	37°
	41°	10.45(4)	10.47(3)	10.42(3)	
Experiments B	36°	10.85(2)	10.83(3)	10.77(3)	38.5°
	41°	10.62(3)	10.81(3)	10.69(2)	
Experiments C	36°	10.62(2)	10.55(1)	10.48(2)	37.7°
	41°	10.57(3)	10.47(3)	10.49(2)	

[a] Numberical values are of $T'_\perp$ (approx) for $I(7,6)$ expressed in gauss. The accuracy of measurement is about $\pm$ 0.08. $T'_\perp$ (approx.) is inversely proportional to order parameter.

[b] The number in parenthesis indicates the number of determinations averaged. A separate plate of cells was used for each determination.

Table II shows that there are small differences in fatty acid composition when normal and transformed CEF, or membranes prepared from them, are compared. These results are in agreement with published fatty acid compositions of CEF (RSV) (11).

Discussion

One characteristic motion of membrane lipids is rapid isomerizations (rate of isomerization $>10^9$ sec^{-1}) of fatty acid chains. Measurement of this motion using fatty acid spin labels shows that there is a gradient of chain flexibility, increasing toward the terminal methyl group (9). Because of the gradient of motion, a quantitative estimate of the mobility of membrane proteins in fluid lipid bilayers cannot be made directly from fatty acid spin label data. It is reasonable to expect, however, that a change in chain flexibility, detected with these spin labels, could result in a change in mobility of membrane proteins. The comparison of the fluid lipid regions of red blood cell membranes and chick cell membranes shown in Figure 1 demonstrates that these labels are quite sensitive to differences in membrane fluidity. It is of interest that lectin-induced receptor aggregation has been observed on trypsin-treated red blood cell membranes (33) as well as on transformed chick cell membranes, although the fluidity of the two types of membrane is quite different. (Trypsin treatment produces only very small, or negligible changes in red blood cell membranes (B.J. Gaffney, unpublished)).

In our experiments, we have been unable to detect a clear difference in fluidity of the lipid regions of the membranes of intact normal and RSV-transformed CEF. Several types of membrane changes might have been detected with the spin label fatty acids. Although changes in the level of unsaturated fatty acid produce changes in order parameter (22) it is doubtful that the small

TABLE II
Fatty Acid Methyl Esters of Whole
Chick Embryo Fibroblasts and Plasma Membrane
Preparations

Cells

	Normal 41°C % of Total	Wild Type RSV 41°C % of Total
16:0	21.4	20.0
16:1	4.1	4.0
18:0	19.5	18.6
18:1	32.7	35.3
18:2	10.0	12.0
20:4	12.3	9.6

Plasma Membranes

	Normal		RSV	
	36°C % of Total	41°C % of Total	36°C % of Total	41°C % of Total
16:0	25.6	24.2	23.2	21.8
16:1	3.6	3.5	5.1	4.5
18:0	20.0	19.7	19.0	17.5
18:1	29.4	30.5	32.0	31.0
18:2	10.8	10.8	13.6	13.9
20:4	10.6	11.3	8.2	11.4

* Average of 2 independent determinations. All
deviations were less than 0.6%.

differences shown in Table II would have been detected in our measurements. Changes in the ratio of cholesterol to phospholipid of $\sim$ 5% might have been detected (23). The possible effects of the membrane proteins on the measured lipid fluidity is harder to gauge. However, from a plot of   % rhodopsin in rhodopsin-lipid recombitants vs. order parameter (24), we may estimate that a change of 15-20% by weight in the fraction of protein in the membrane could be detected as a change in lipid fluidity. In addition, small changes in the fluidity of the RBC membrane, which cannot have arisen from any of the above sources, have been detected recently by spin label measurements. These changes result from the interaction of prostaglandins E_1 and E_2 (5) and carbamoylcholine (6) with the red blood cell membranes. In both cases, the changes were reversible and arose from very low concentrations of added reagent (10^{-12}M for PGE_1 and E_2 and 10^{-6}M for carbamoyl choline).
    The fact that we find no differences in the average lipid fluidity of normal and transformed CEF has two implications. Either 1) substantial changes in the average properties of fluid lipid regions of membranes are not associated with viral-transformation or 2) more than one change is involved in CEF cells in such a way that changes in lipid fluidity of equal magnitude but opposite sign result from two or more opposing effects. The complexity of the relation between membrane fluidity and transformation is in fact quite well illustrated by recent results with fluorescent labels (34,35). It was found that increased mobility of fluorescent Con A receptors accompanied transformation of fibroblasts in cultures while decreased mobility characterized the state of Con A receptors on transformed cells in suspension (34). In addition, when normal lymphocytes and malignant lymphoma cells were compared, fluorescent Con A showed a decreased mobility on the lymphoma cells (34) while fluorescent

111

probes in lipid regions of the same cells indicated that the transformed lymphocytes had enhanced lipid mobility (35).

A possible mechanism by which lipids might influence membrane protein mobility, which is not inconsistent with any of the above observations, is motion of the proteins along boundaries of solid and fluid lipid phases. This possibility is suggested by the finding of Petit and Edidin (18) that there is an acceleration in mobility of surface antigens between $20°$ and $15°$ C in mouse-human heterokaryons. Another possibility, which is not necessarily distinct from the one above, is that changes in the fluidity of small localized areas of the cell membrane may be an important feature of transformation. Comparison of the fluidity of virus lipid membranes with the membranes of the host cells suggests that specialized patches of cell membranes, differing in lipid fluidity, may indeed exist (36).

References

1. "The Molecular Biology of Tumor Viruses," ed. J. Tooze, Cold Spring Harbor Laboratory, 1973, Chapter 3.
2. G.L. Nicolson, Nature New Biol. 233, 244 (1971).
3. K.D. Noonan and M.M. Burger, J. Cell Biol. 59, 134 (1973).
4. J.Z. Rosenblith, T.E. Ukena, H.H. Yin, R.D. Berlin and M.J. Karnovsky, Proc. Nat. Acad. Sci. 70, 1625 (1973).
5. P.G. Kury, P.W. Ramwell and H.M. McConnell, Biochem. Biophys. Res. Commun., in press (1974).
6. W.H. Heustis and H.M. McConnell, Biochem. Biophys. Res. Commun., in press (1974).
7. H.M. McConnell and B.G. McFarland, Quart Rev. Biophys. 3, 91 (1970).
8. W.L. Hubbell and H.M. McConnell, J. Am. Chem. Soc. 93, 314 (1971).
9. B.J. Gaffney and H.M. McConnell, J. Mag. Res. in press (1972).
10. S. Kawai and H. Hanafusa, Virology 46, 470 (1971).
11. T.M. Yau and M.J. Weber, Biochem. Biophys. Res. Commun. 49, 114 (1972).
12. M.M. Burger and G.S. Martin, Nature New Biol. 237, 9 (1972).
13. M. Kapeller and F. Doljanski, Nature 235, 184 (1972).
14. J.M. Lehman and J.R. Sheppard, Virology, 49, 339 (1972).
15. M.M. Poo and R.A. Cone, Nature 247, 438 (1974).
16. R.A. Cone, Nature New Biology 236, 39 (1972).
17. C.D. Linden, K.L. Wright, H.M. McConnell and C.F. Fox, Proc. Nat. Acad. Sci. 70, 2271 (1973).
18. V.A. Petit and M. Edidin, Science in press (1974).
19. I. Yahara and G.M. Edelman, Proc. Nat. Acad. Sci. 69, 608 (1972).

20. M.C. Raff and S. DePetris, Fed. Proc. 32, 48 (1973).

21. E.J. Shimshick and H.M. McConnell, Biochem. 12, 2351 (1973)

22. S. Rottem, W.L. Hubbell, L. Hayflick, and H.M. McConnell, Biochim. Biophys. Acta 219, 104 (1970).

23. E.J. Shimshick and H.M. McConnell, Biochem. Biophys. Res. Commun. 53, 446 (1973).

24. K. Hong and W.L. Hubbell, Proc. Nat. Acad. Sci. 69, 2617 (1972).

25. A. Rein and H. Rubin, Ex. Cell Res. 49, 666 (1968).

26. B.J. Gaffney, in Spin Labeling: Theory and Applications, ed. L.J. Berliner, Academic Press, in press.

27. D.M. Brunette and J.E. Till, J. Membrane Biol. 5, 215 (1971).

28. R. Dulbecco and M. Vogt, J. Exp. Med. 99, 167 (1954).

29. T. Smith, T. Brooks and H. White, Lipids, 4 31 (1969).

30. G.Y. Sun and L.A. Horrocks, J. Lipid Res. 10, 153 (1969).

31. P. Devaux and H.M. McConnell, J. Am. Chem. Soc. 94, 4475 (1972).

32. C.J. Scandella, P. Devaux and H.M. McConnell, Proc. Nat. Acad. Sci. 69, 2056 (1972).

33. G.L. Nicolson, in Membrane Research, ed. C.F. Fox, Academic Press, New York (1972).

34. M. Inbar, M. Shinitzky and L. Sachs, J. Mol. Biol. 81, 245 (1973).

35. M. Shinitzky and M. Inbar, J. Mol. Biol., in press (1974).

36. B. Sefton and B. Gaffney, manuscript in preparation.

Acknowledgements

We are grateful to Dr. R. Holm, M.I.T., for making the E-9 spectrometer available to us.

This work has been supported by the NIH under grant 9-R01-CA14142-11 (to P.W. Robbins) and under grant GB33501X-1 (to H.M. McConnell) and by a faculty award from a NIH Biomedical Sciences Support Grant to JHU (to BJG).

TRANSFER RNAs OF ROUS SARCOMA VIRUS AND THE INITIATION
OF DNA SYNTHESIS BY VIRAL REVERSE TRANSCRIPTASE

B. Cordell-Stewart, J.M. Taylor, W. Rohde,
H.M. Goodman and J.M. Bishop

Departments of Microbiology and Biochemistry & Biophysics
University of California, San Francisco, California 94143

Summary

A 4S RNA which serves as primer for the initiation of
DNA synthesis by reverse transcriptase has been isolated from
the 70S RNA of Rous sarcoma virus. This primer has structural characteristics of tRNA and can be aminoacylated with
methionine, but its chromatographic behavior and features of
its nucleotide sequence indicate that it is not one of the
four identified forms of avian met-tRNA. Primer is bound
to the viral genome at specific sites (ca. 2-4 per genome)
which cannot be occupied by other 4S RNAs; annealing of
purified primer to these sites results in a functional complex of template-primer. A 4S RNA structurally similar to
primer has been isolated from normal vertebrate cells by
Dahlberg and coworkers. The cellular RNA is a trp-tRNA by
all available criteria. We cannot presently reconcile these
observations with the functional and chromatographic properties of primer isolated from viral RNA.

Introduction

Virions of avian sarcoma viruses contain three distinct
populations of 4S RNA. The most abundant and heterogeneous
group is found free of the 60-70S viral genome (free 4S RNA),
whereas the other two populations are physically associated
with the genome (70S-a 4S RNA) and are released by denaturation of the 70S RNA complex (Table 1). The bulk of 70S-a
4S RNA (ca. 85%) can be removed from the viral genome by
heating at 63° in buffers containing low concentrations of
cation (Faras et al., 1973; Dahlberg et al.,1974; Canaani
and Duesberg, 1972); the remainder is released by heating at
higher temperatures (eg., 80°, Dahlberg et al., 1974; Canaani
and Duesberg, 1972) or by treatment with dimethylfulfoxide
(Dahlberg et al., 1974). We have shown previously that the
80°-fraction of 70S-a 4S RNA is a structurally homogeneous
RNA which serves as primer for the initiation of DNA synthesis in vitro when 70SRNA is used as template for reverse
transcriptase (Dahlberg et al., 1974; Faras et al., 1973).

The present communication summarizes current data regarding the nature of this primer RNA and explores the extent of its structural and functional relationship to a similar RNA isolated from normal cells by other investigators (Sawyer et al., 1974).

Results

Electrophoretic mobility, composition and nucleotide sequence of primer RNA.

Sawyer and Dahlberg have demonstrated that two-dimensional electrophoresis in polyacrylamide gels fractionates the 4S RNAs of Rous sarcoma virus (RSV) into discrete and apparently homogeneous species (Sawyer and Dahlberg, 1973). In collaboration with these investigators, we used this technique to analyze each of the three populations of RSV 4S RNA defined above (i.e., free 4S RNA, and the 63° and 80° fractions of 70S-a 4S RNA) and to purify the 4S primer RNA (Dahlberg et al., 1974). Two-dimensional electrophoresis of free 4S RNA resolves 10-15 different RNA species (see Sawyer and Dahlberg, 1973 and Fig. 1a), most of which are recovered at an average of 1 copy per virion (or per viral genome). However, the rapidly migrating species denoted "Spot 1 RNA" is present in molar excesses of 10-30 (Table 1; Sawyer and Dahlberg have reported somewhat lower values (6-8); Sawyer and Dahlberg, 1973). The 63° fraction of 70S-a 4S RNA is also relatively complex, containing a minimum of 8 discrete species of 4S RNA (Fig. 1b). "Spot 1 RNA" is recovered in this population at 0.5 molecules per viral genome. The 80° fraction of 70S-a 4S RNA contains only "Spot 1 RNA" at an average of 2-4 molecules per viral genome (Fig. 1c). (We have no evidence regarding the distribution of these 4S RNAs among the molecules of 70S RNA.)

We analyzed the nucleotide composition and sequence of the 80° fraction of 70S-a 4S RNA and "Spot 1 RNA" obtained from both free 4S RNA and the 63° fraction of 70S-a 4S RNA; the results indicated that these various RNAs are very similar if not identical and led us to conclude that "Spot 1 RNA" from all the sources analyzed is the 4S primer of RSV (Dahlberg et al., 1974; Faras et al., 1974).

Purified primer (80° fraction of 70S-a 4S RNA) and "Spot 1 RNA" have compositional and structural features of tRNA. These features include size (75 nucleotides), the 3' ($pCpCpA_{OH}$) and 5' (pGp) termini, and the presence of at least ten different modified nucleosides (Faras et al., 1974). The molecule is distinguished by the absence of ribothymidine

TABLE 1

Transfer RNAs of RSV and the Initiation of DNA
Synthesis by Reverse Transcriptase

	80° 70S-a	63°4S	63° Spot 1	Free 4S	Free 4S Spot 1	Cell 4S Spot 1
Amino acid[1]	met no trp	met no trp	nt	met little trp	nt	trp no met
Sequence data[2]	t-RNA	t-RNAs	t-RNA	t-RNAs	t-RNA	trp t-RNA
Proportion of total[3]	100%		5%		25%	3-5%
Molecules per virion[4]	2-4	10-20	0.5	80-120	20-30	-
Reconstitution of template activity[5]	+	+	nt	+	+	nt

[1] See Table 2; J. Dahlberg, personal communication.
[2] Sawyer and Dahlberg(1973); J. Dahlberg, personal communication, Faras et al.(1974).
[3] See Figure 1; also, Dahlberg et al., (1974); Sawyer and Dahlberg (1973).
[4] Relative yields were obtained by determining the radioactivity present in each species of RNA resolved by 2-dimensional gel electrophoresis. These counts were normalized for chain length (cpm/nucleotide) and the number of molecules of each RNA or RNA fraction was calculated relative to 1 copy of 70S RNA per virion.
[5] See Table 7.

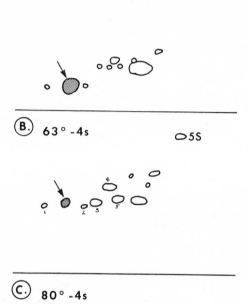

FIGURE 1

The Low Molecular Weight RNAs of RSV: Fractionation by two

Dimensional Electrophoresis in Polyacrylamide Gels

^{32}P-labeled low molecular weight RNAs were isolated
from purified RSV and RSV 70S, then analysed by two-dimensional electrophoresis in gels of 10% (first dimension) and
20% polyacrylamide. Each panel in the figure is a diagram
of an autoradiogram prepared with the 20% gel after the
conclusion of the second dimension of electrophoresis. 63°
and 80° 4S RNA fractions were prepared by sequential denaturation of 70S RNA in 0.01M EDTA 0.02M Tris:HCl, pH 7.4
(Dahlberg et al., 1974). The arrows indicate the 4S RNA
identified as primer (Spot 1) (Dahlberg et al., 1974).

120

(as are a number of eukaryotic tRNAs, see Petrissant, 1973; Simsek et al., 1973; Marcu et al., 1973) and the presence of the nucleotide sequence GpψpψpCpGp in place of the oligo-nucleotide GprTpψpCpGp characteristic of many tRNAs. Very few eukaryotic tRNAs have been analysed in this manner; consequently, it is not possible to determine whether the nucleotide sequence of RSV primer is in any way unusual.

Dahlberg and coworkers have identified a "Spot 1 RNA" in the 4S RNA from a number of vertebrate animals, including chickens, ducks, rats and humans (Sawyer et al., 1974). They have determined the complete nucleotide sequence of this RNA from chicken; the RNA has the anticodon for tryptophan and shares approximately 60% of its sequence with that of trp-tRNA from yeast (personal communication). Less complete analyses of "Spot 1 RNA" from both free 4S RNA and 70S-a RNA of RSV suggests that these RNAs have a nucleotide sequence similar to that of cellular "Spot 1 RNA" (Faras et al., 1974); consequently, Dahlberg and his colleagues have concluded that the primer of RSV is a trp-tRNA (personal communication, and Proceedings of the 1974 Cold Spring Harbor Symposium on Tumor Viruses).

Aminoacylation of viral 4S RNAs.

The three classes of virus-associated 4S RNA have been identified as tRNAs by the functional criterion of amino-acylation. The free 4S RNAs of RSV and AMV possess a wide spectrum of aminoacid acceptor activity (Trávnicek, 1969; Erikson and Erikson, 1970; Wang et al., 1973) illustrated by the data summarized in column one of Table 2. The highest levels of activity are found with methionine, lysine, leucine, proline and histidine. When compared to the tRNA activity found in either uninfected or transformed cells, the virion free 4S RNA cannot be considered a random collection of host cell tRNAs. All available data indicate an enrichment for several species; methionine is most prominent, followed by lysine and histidine.

Native 70S RNA does not accept aminoacids (Table 3; Erikson and Erikson, 1970), but the 4S RNAs released from the viral genome by denaturation can be acylated with a variety of aminoacids (Table 2; Rosenthal and Zamecnik,1973). These results suggest that the binding of 4S RNA to the genome imposes conformational or steric hindrances which prohibit functional activity; we reached a similar conclusion from the fact that the 4S RNAs in native 70S RNA are not substrates for tRNA nucleotidyl transferase (Faras et al., 1973).

TABLE 2

Acylation of RSV 4S RNAs with Amino Acids[1]

Amino Acid	Acceptance by 4S RNAs (pmole amino acid/pmole RNA)				
	Free 4S SR-RSV[2]	AMV[3]	Total 70S-a 4S AMV[4]	(63°) 4S PR-RSV(C)	(80°) 4S PR-RSV(C)
Alanine	.002	.003	–	.002	–
Arginine	.002	.010	–	.006	–
Asparagine	.002	.002		.005	–
Aspartic	.002	.002		–	–
Cystine	nt	nt		–	–
Glutamic	.002	–		.001	–
Glutamine	–	.004		–	–
Glycine	nt	.008	.001	.002	–
Histidine	.008	.008	.013	.019	–
Hydroxyproline	nt	nt		–	–
Isoleucine	.005	.007		.005	–
Leucine	.014	.003	–	.004	–
Lysine	.010	.011	.040	.035	–
Methionine	.012	.024		.091	.226
Phenylalanine	.001	.002	.006	.003	–
Proline	.008	.018	.001	.002	–
Serine	–	–	.015	.027	–
Threonine	.004	.005	.015	.007	–
Tryptophan	.003[5]	nt		–	–
Tyrosine	–	–		–	–
Valine	.006	.001		.040	–
TOTAL	.081	.109		.249	.226

[1] – denotes background, i.e., <0.0005 pmole/pmole.
Primer was separated from the remainder of 70S-a 4S RNA by sequential denaturation at 63° and 80° (Dahlberg et al., 1974). This produces two fractions of 4S RNA: 63°- 4S; (released at 63°) and purified primer (released at 80°). Purity of the primer was documented by analysis of the oligonucleotides released by hydrolysis with T_1 RNase (Faras et al., 1974). Acylation with [^{3}H]amino acids was carried out according to conventional techniques using aminoacyl tRNA synthetases prepared from chicken liver (a gift of E. Penhoet).
[2] Data from Gallagher and Gallo (1973).
[3] Data from Wang et al.(1973) except tryptophan value.
[4] Data from Rosenthal and Zamecnik (1973).
[5] This value was determined inadependently in our laboratory.

TABLE 3

Aminoacylation of Native and Heat Dissociated RSV 70S RNA

	Incorporation of ^{3}H-Protein Hydrolysate cpm
no RNA	546
70S-untreated	575
70S-63° only	1403
70S-63° + 80°	1769

Samples of 10 µg of purified RSV 70S RNA in 0.01M EDTA 0.02M Tris:HCl, pH 7.4 were partially or completely dissociated by heating at 63° for 3 minutes then 80° for 1 minute, followed by a fast cool in ice. The RNA was added directly to a conventional t-RNA acylation reaction mixture containing radiolabeled amino acids from a reconstituted protein hydrolysate.

The number of acceptor activities in the 63° fraction of 70S-a 4S RNA corresponds approximately to the number of RNA species resolved by two-dimensional electrophoresis; appreciable acylation is found with methionine, valine, lysine, histidine, threonine and serine (Table 2). Similar results have been reported for the total 70S-a 4S RNA of AMV (Rosenthal and Zamecnik, 1973; Elder and Smith, 1974; data summarized in Table 2).

The 80° fraction of 70S-a 4S RNA aminoacylates only with methionine (Table 2); this observation conforms to structural analyses which indicate that the RNA is a homogeneous population (Faras et al., 1974). By contrast, the primer-like RNA isolated from uninfected chicken cells ("Spot 1 RNA"; see Sawyer et al., 1974) accepts tryptophan, in accord with the nucleotide sequence of the RNA (personal communication from J. Dahlberg). This is an unexpected result because "Spot 1 RNA" is a prominent constituent of both free and 70S-a 4S RNA, yet tryptophan acceptance by these RNAs is either low or undetectable under conditions which permit ample charging of cellular trp-tRNA. Both sets of 4S RNA have high levels of methionine acceptance (Tables 2 and 4). Nevertheless, published data clearly indicate that cellular spot 1 RNA (which accepts tryptophan) and primer RNA (the 80° fraction, which accepts methionine) are structurally similar, although minor differences in nucleoside modification may exist (Faras et al., 1974). There are several possible explanations for these discrepancies. 1) The charging of primer with methionine is an experimental artifact; however, Faras has independently obtained similar results (personal communication), and our available data indicate that the charge by methionine is a genuine aminoacylation. 2) Preparative manipulations of primer induce conformational changes which cause the molecule to accept an "incorrect" aminoacid; we have no definitive evidence against this explanation. 3) The process by which cellular trp-tRNA is incorporated into a complex with the viral genome alters the specificity of aminoacylation; it is impossible to properly evaluate this possibility because we know virtually nothing of the mechanisms which facilitate the binding of tRNA to the viral genome. 4) The isolated primer is inactivated for tryptophan aminoacylation and is contaminated with a met-accepting tRNA; however, primer appears to be homogeneous when subjected to compositional analyses (Dahlberg et al., 1974; Faras et al., 1974) and chromatography on "reversed phase" columns (see below).

TABLE 4

Comparison of Aminoacylation Versus Primer Mass
in the Three Fractions of RSV 4S RNAs

	Proportion of Total 4S RNA as Spot 1	Proportion of Total Amino Acid Acceptor Activity	
		met	trp
Free 4S	20-30%	23%[+]	2%
63° 4S	5%	36%[+]	< 2%
80° 4S	100%	100%	< 2%

[+] These values include all the isoaccepting species of
met-tRNA present in the individual fractions of viral RNA.

Chromatography of viral tRNAs.

Chromatography by the "reversed phase" procedure (RPC) permits the separation and identification of isoaccepting species of tRNAs (Pearson et al., 1971). We have used RPC to compare aminoacylated primer to other forms of tRNA in RSV (Fig. 2). Both the free 4S and the 63° fraction of 70S-a 4S RNA contain four separable species of met-tRNA which cochromatograph with met-tRNAs from chick cells (form I is an initiator tRNA, the other three forms donate methionine to internal peptide bonds; Elder and Smith, 1973; and personal communication from A.J. Faras). The two sets of viral RNA contain different proportions of the various met-tRNAs; form IV predominates in the free 4S RNA, whereas all four forms are present in major amounts in the 63° fraction of 70S-a 4S RNA. These results confirm and extend recently published data from other investigators (Wang et al., 1973; Elder and Smith, 1973; Elder and Smith, Gallagher and Gallo, 1973).

Aminoacylated and radiolabeled primer chromatographs as a single species of RNA, usually separable from the four identified forms of cellular met-tRNA (Fig. 2c). Cellular trp-tRNA (Fig. 2f) and the trp-tRNA in RSV free 4S RNA (Fig. 2e) chromatograph similarly and are well separated from primer despite the apparent structural identity between primer (i.e., 80° fraction of 70S-a 4S RNA) and cellular trp-tRNA. "Spot 1 RNA" isolated from free RSV 4S RNA by two dimensional electrophoresis chromatographs in a position similar to that of primer purified from 70S-a 4S RNA (Fig. 2d).

Reannealing of primer to viral genome.

Canaani and Duesberg have demonstrated that the primer molecules in 70S-a 4S RNA can be reannealed in a functional configuration to high molecular weight subunits of the RSV genome (Canaani and Duesberg, 1972). We have repeated these experiments with primer purified by sequential denaturation (80° fraction), and with "Spot 1 RNA" isolated by two-dimensional electrophoresis from both free 4S RNA of RSV and the 63° fraction of 70S-a 4S RNA. Reassociation of radiolabeled 4S RNA with viral genome is detected and quantified by electrophoresis in polyacrylamide gels (manuscript in preparation). Primer RNA from each of the sources just described anneals completely to viral genome under conditions which do not permit any other form of viral or cellular tRNA to anneal (Table 5). The reannealed complexes of template-

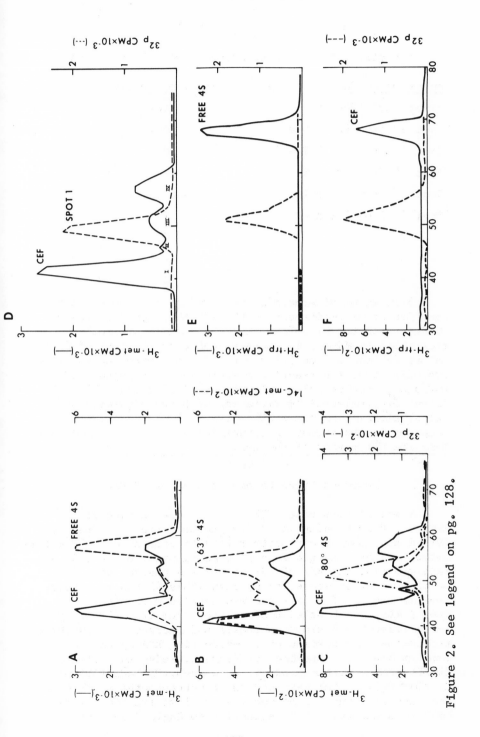

Figure 2. See legend on pg. 128.

127

TABLE 5

Annealing of 4S RNAs to Subunits of the RSV Genome

Source of 4S RNA	Extent of Annealing
Primer (80° 70S-a 4S)	100%
"Spot 1" from 80° 70S-a 4S	100%
"Spot 1" from 63° 70S-a 4S	100%
"Spot 1" from free 4S	100%
"Spot 1" from chick 4S	Not tested
"Spot 2" from 70S-a 4S	0
3	0
4	0
5	0

3-30 µg/ml of subunits, prepared by rate-zonal centrifugation of heat-dissociated RSV 70S RNA, and ^{32}P-labeled
4S RNA, isolated by 2-dimensional gel electrophoresis
(described in Figure 1), were annealed in 10-20 µl of 0.6M
Na^+ at 74° for 2.5 to 3.5 hours. To determine the extent of
hybridization, the reaction mixture was electrophoresed
through a stacked polyacrylamide gel (2.25% and 10%). The
amount of labeled-4S RNA migrating with the high-molecular
weight subunits in the 2.25% portion of the gel was then
compared to the amount of unannealed 4S RNA electrophoresed
into the 10% region of the gel.

FIGURE 2

Reversed Phase Chromatography of 4S RNAs

Abbreviations used: CEF, chick t-RNA charged with
either [^{3}H]methionine or [^{3}H]tryptophan; Free 4S, free 4S
RNA of RSV charged with either [^{3}H]-tryptophan or [^{14}C]methi-
onine; 63° 4S, charged with [^{14}C]methionine; 80° 4S, 4S RNA
released from 70S RNA of RSV at 80° and charged with[^{14}C]
methionine or unlabeled tryptophan (this 4S RNA is a homogeneous sample of primer and was labeled with ^{32}P).
Spot 1 RNA, ^{32}P-labeled RNA isolated from free RSV 4S by
2-dimensional gel electrophoresis and charged with unlabeled
methionine. Separation of iso-accepting RNA species was
done on RPC-5 as described by Pearson, Weiss and Kelmers
(1971) with a linear gradient of 0.45-0.70M NaCl (100 ml)
in 10mM $MgCl_2$, 10mM NaAc and 2mM β-mercaptoethanol pH 4.5.
The flow rate was 0.5 ml per minute; 1 ml fractions were
collected and counted in Aquasol (New England Nuclear).

primer denature sharply with a T_m of 70° in 0.01M EDTA-0.02M Tris-HCl (data not shown); similar results were obtained previously with native template-primer complexes (Dahlberg et al., 1974; Canaani and Duesberg, 1972). Analysis of the kinetics of annealing indicates that each genome subunit contains an average of 1-2 binding sites for primer (manuscript in preparation); these results conform to the calculated yields of primer purified from 70S RNA by several techniques (Dahlberg et al., 1974; Sawyer and Dahlberg,1973).

Annealing of primer to viral genome can be competitively inhibited by excesses of tRNA from normal cells of ducks, chickens and mice, but not by tRNAs from E. coli or Baker's yeast (Table 6). These results substantiate Dahlberg's conclusion that vertebrate (and only vertebrate) cells contain 4S RNA identical (or very similar) to the primer found in RSV and its natural host (Sawyer et al., 1974). We have yet to identify the individual species of tRNA responsible for competition.

High molecular weight subunits isolated from denatured 70S RNA of RSV have very little activity as template for DNA synthesis by reverse transcriptase of RSV (Canaani and Duesberg, 1972; Faras et al., 1972); annealing of primer (80° fraction of 70S-a 4S RNA) to subunits reconstitutes template activity (Table 7). Similar results were also obtained with either the total 4S RNA of chicken cells or free 4S RNA of RSV, presumably because both populations of RNA contain primer molecules (Dahlberg et al., 1974; Fig. 1 and Table 2). Canaani and Duesberg were unable to reconstitute template activity with cellular 4S RNA; our at least limited success (maximum of 28%) was probably due to the use of larger amounts of cellular RNA in order to obtain an adequate excess of primer molecules.

Aminoacylation of primer with methionine prior to performing the annealing with subunits substantially reduces the extent to which template activity is reconstituted (Table 8). These provisional data implicate the 3'-terminus which is subject to acylation in the initiation of DNA synthesis. The blockage of initiation is incomplete; this may reflect: (a) failure to completely acylate the primer RNA, (b) deacylation during the subsequent experimental manipulations (these were carried out at pH 4-6 in order to stabilize the aminoacyl bond), or (c) heterogeneity of the primer population.

TABLE 6

Annealing of 4S RNA to Subunits of the RSV genome:

Competition Between Primer and Other 4S RNAs

Test RNA	Competitor	Extent of Annealing[+]
A. Primer (80°70S-a 4S)	None	51%
	Duck 4S	10%
	Yeast 4S	45%
B. Primer	None	69%
	63° 70S-a 4S	18%
	Mouse 4S	12%
	E. coli 4S	68%
C. "Spot 1" from Free 4S of RSV	None	46%
	Chick 4S	14%
	Mouse 4S	11%
	Hamster 4S	13%
	E. coli 4S	51%

[+] Data for competitor RNAs are taken from points at or near maximum competition. The conditions and assay of annealing were identical to those described in Table 5. Each un-labeled 4S RNA competitor was present in a 1000X excess of the ^{32}P-80° 4S primer. The reactions were deliberately limited so that annealing was incomplete; this provides a sensitive index for competition.

TABLE 7

Reconstitution of Template Activity with

RSV Subunits and 4S RNAs

	DNA Synthesis with RSV DNA Polymerase
70S RNA	100%
35S subunits	6%
80° 4S[a]	4%
63° 4S[b]	8%
Free 4S[c]	9%
CEF 4S[d]	3%
35S annealed w/80° 4S[a]	73%
35S annealed w/63° 4S[b]	55%
35S annealed w/free 4S[c]	84%
35S annealed w/CEF 4S[d]	28%

35S subunits were obtained by dissociating purified 70S RNA at 100°C for 1 minute in 0.01M EDTA 0.02M Tris:HCl, pH 7.4, then fractionating the high and low molecular weight RNAs by velocity centrifugation. 2 μg subunits were annealed with the indicated amounts of each 4S RNA in 50 μl of 6X SSC (1X SSC = 0.15M NaCl 0.015M sodium-citrate) after the procedure described by Canaani & Duesberg (1972). Template activity of 1.0 μg of RNA [in 50 μl of 0.1M Tris pH 8.1, 0.01M $MgCl_2$ with 2% β-mercaptoethanol] was determined by a 1.5 hr incubation in a standard polymerase assay (Faras et al., 1972).

[a] 0.2 μg 4S

[b] 1.0 μg 4S

[c] 1.0 μg 4S

[d] 2.0 μg 4S

TABLE   8

Reconstitution of Template Activity with

RSV Subunits and 4S Primer

	DNA Synthesis with RSV DNA Polymerase
Control (70S RNA)	100%
Subunits (35S)	3%
Primer (4S)	1%
Subunits annealed with primer	70%
Subunits annealed with met-primer	25%
Subunits annealed with deacylated met primer	58%

Purified primer was charged with methionine as in Table 2.   A portion was then deacylated by incubation at pH 8.   RNAs were annealed and tested as template for purified DNA polymerase of RSV as described in Table 7.

Conclusion

Transcription of DNA in vitro from the 70S RNA of RSV initiates on the 3'-terminus of an RNA with structural and functional features of tRNA. This RNA is structurally similar to a 4S RNA isolated from uninfected vertebrate cells (Sawyer et al., 1974), but according to available data the viral and cellular RNAs can be acylated by different amino acids (Table 1). We cannot presently explain this discrepancy.

The primer for viral DNA synthesis in the infected cell has not been identified, but several lines of evidence indicate that the synthesis of DNA in vitro by reverse transcriptase with 70S RNA as template does not accurately duplicate events in the infected cell (Bishop et al., 1974). It is possible that initiation in vivo may occur by presently unrecognized mechanisms.

Reverse transcriptase can copy the entire genome of RSV into short chains (ca. 100-200 nucleotides) of DNA; this fact implies the existence of multiple initiation sites (as many as several hundred) on each molecule of viral RNA, yet data presented here and elsewhere (Dahlberg et al., 1974; Sawyer and Dahlberg, 1973) indicate that each genome contains on the average only 2-4 primer molecules. We cannot presently reconcile these apparently conflicting observations because nothing is known of either the distribution of primer molecules among the population of template RNA or the topography of individual viral genomes. We are now attempting to map the positions of binding sites for primer RNA on subunits of 70S RNA.

There is presently no indication that any of the other tRNAs associated with RSV have functions in the viral life cycle, and the available evidence suggests that these RNAs may be superfluous to viral growth. For example, our data indicate that only primer RNA is specifically bound to viral genome by reasonably stable regions of secondary structure; the other 70S-a 4S RNAs dissociate from the genome at relatively low temperatures (Dahlberg et al., 1974; Canaani and Duesberg, 1972) and cannot be reannealed to the genome under conditions favorable for the reformation of RNA-RNA duplexes (Table 5). The mechanisms by which these 4S RNAs (and primer, as well) are reproducibly included in the virions and complexed with the genome remain to be elucidated.

133

Acknowledgements

Some of the work described here evolved from a pleasant collaboration with Drs. J.E. Dahlberg, R. Sawyer and F. Harada; in particular, we acknowledge that the analyses shown in Figure 1 are similar to previous work carried out in their laboratory, but repeated by us and illustrated here for the sake of clarity. Work in the authors' laboratories is supported by grants from the American Cancer Society VC-70A, the USPHS AI 08864, CA 12705, CA 14026, and Contract No. NO1 CP 33293 within The Virus Cancer Program of the National Cancer Institute. J.M.T. acknowledges the support by Senior Dernham Fellowship (D-201) of the American Cancer Society, California Division.

References

1. A.J. Faras, A.C. Garapin, W.E. Levinson, J.M. Bishop, and H.M. Goodman, J. Virol. 12, 334, (1973).
2. J.E. Dahlberg, R.C. Sawyer, J.M. Taylor, A.J. Faras, W.E. Levinson, H.M. Goodman, and J.M. Bishop, J. Virol., in press, (1974).
3. E. Canaani, and P. Duesberg, J. Virol. 10, 23, (1972).
4. A.J. Faras, J.M. Taylor, W.E. Levinson, H.M. Goodman, and J.M. Bishop, J. Mol. Biol. 79, 163, (1973).
5. E. Erikson, and R.L. Erikson, J. Virol. 8, 254, (1971).
6. R.C. Sawyer, and J.E. Dahlberg, J. Virol. 12, 1226, (1973).
7. R.C. Sawyer, F. Harada, and J.E. Dahlberg, J. Virol., in press, (1974).
8. A.J. Faras, J.E. Dahlberg, R.C. Sawyer, F. Harada, J.M. Taylor, W.E. Levinson, J.M. Bishop, and H.M. Goodman, J. Virol., in press(1974).
9. G. Petrissant, Proc. Nat. Acad. Sci. U.S.A. 70, 1046, (1973).
10. M. Simsek, J. Ziegenmeyer, J. Heckman, and V.L. Rajbhandary, Proc. Nat. Acad. Sci. U.S.A. 70, 1041, (1973).
11. K. Marcu, R. Mignery, R. Reszelback, B. Roe, M. Sirover, and B. Dudock, Biochem. Biophys. Res. Commun. 55, 477, (1973).
11a. M. Trávnicek, Biochim. Biophys. Acta. 182, 427, (1969).
12. E. Erikson, and R.L. Erikson, J. Mol. Biol. 52, 387, (1970).
13. S. Wang, R.M. Kotari, M. Taylor, and P. Hung, Nature New Biol. 242, 133, (1973).
14. K.T. Elder, and A.E. Smith, Proc. Nat. Acad. Sci. U.S.A.

70, 2823, (1973).
15. L.J. Rosenthal, and P.C. Zamecnik, Proc. Nat. Acad. Sci. U.S.A. 70, 1184, (1973).
16. A.J. Faras, W.E. Levinson, J.M. Bishop, and H.M. Goodman, Virology, in press, (1973).
17. K.T. Elder, and A.E. Smith, Nature 247, 435, (1974).
18. R.L. Pearson, J.F. Weiss, and A.D. Kelmers, Biochem. Biophys. Acta. 228, 770, (1971).
19. R.E. Gallagher, and R.C. Gallo, J. Virol. 12, 449, (1973).
20. J.M. Bishop, C.T. Deng, A. Faras, H. Goodman, R.Guntaka, W. Levinson, B. Cordell-Stewart, J. Taylor, and H.Varmus To be published in the Proceedings of the Fourth International Symposium on Comparative Leukemia Research, (1974).
21. A.J. Faras, J.M. Taylor, J. McDonnell, W. Levinson, and J.M. Bishop, Biochemistry 11, 2334, (1972).

RECOMBINANTS OF AVIAN RNA TUMOR VIRUSES:
CHARACTERISTICS OF THE VIRION RNA

Peter Duesberg, Karen Beemon, Michael Lai and Peter K. Vogt

Department of Molecular Biology and Virus Laboratory
University of California, Berkeley, California
and
Department of Microbiology, University of Southern
California School of Medicine, Los Angeles, California

Summary

The RNAs of several avian tumor virus recombinants which had inherited their focus forming ability from a sarcoma virus and the host range marker from a leukosis virus were investigated. Electrophoresis and analysis of oligonucleotide fingerprints showed that the cloned sarcoma virus recombinants contained only size class a RNA, although they had acquired a marker which resided on class b RNA in the leukosis virus parent. Class a RNA of different recombinant clones, derived from the same pair of parental viruses and selected for the same biological markers, differed slightly in electrophoretic mobility from each other and from the parental sarcoma virus. Small electrophoretic differences were also observed between the class a RNAs of various strains of avian sarcoma viruses and between class b RNAs of leukosis viruses, but these minor variations in RNA size were not related to the size of recombinant RNAs derived from these viruses.

Recombinants of the same cross and selected for the same pair of markers were also found to have different fingerprints of RNase T1 resistant oligonucleotides. The average complexity of the 60-70S RNA prepared from wild type sarcoma viruses was estimated to correspond to 2.7×10^6 daltons, suggesting that the genome of RNA tumor viruses is polyploid.

All these observations led us to propose that recombination among avian tumor viruses occurs by crossing over between homologous pieces of nucleic acid.

Introduction

Nondefective avian sarcoma viruses can undergo high frequency genetic recombination with avian leukosis viruses (1,2,3). Since the 60-70S RNA of avian tumor viruses consists of several pieces (4), it appeared likely that this recombination represented reassortment of markers situated on different genome subunits. However preliminary electrophoretic analysis of the RNA from a recombinant between PR-B sarcoma and RAV-3 leukosis virus led us to propose that recombinants between avian tumor viruses originate from crossing-over (5,6).

This proposal was based on the following argument: Cloned nondefective sarcoma viruses contain only 30-40S RNA of size class a. The 30-40S pieces of leukosis viruses are of the smaller size class b. In a cross between PR-B sarcoma and RAV-3 leukosis virus, the recombinant is selected for the focus forming ability of the sarcoma virus linked to the host range marker of the leukosis virus, thus combining markers which on parental RNAs are situated on size a and b molecules respectively. However, the RNA of the PR-B x RAV-3 recombinant contained only class a pieces, and thus the leukosis-derived marker must have become incorporated into class a RNA.

To distinguish better between reassortment and crossing over, additional recombinants which had inherited the focus forming marker from a sarcoma and the host range marker from a leukosis virus were investigated. The RNAs of recombinants and of parental viruses were compared with respect to electrophoretic mobility, RNase T1 fingerprint pattern, and genetic complexity. The results favor crossing over and suggest that the genome of RNA tumor viruses may be polyploid.

Results

The RNAs of several recombinant sarcoma viruses obtained from different host range crosses.

Figure 1A shows an electropherogram of heat dissociated 60-70S RNA from a recombinant between the focus forming marker of PR-A and the host range marker of RAV-2. This recombinant contained only 30-40S RNA of class a, which coincided with the class a RNA of the PR-C standard (the latter showed a class b component as well, probably representing a transformation-defective segregant which had formed during nonclonal passage of this virus stock) (5,7,8). The RNA of a recombinant between PR-B focus for-

mation and RAV-1 host range was co-electrophoresed with the
PR-A x RAV-2 recombinant of Figure 1A, and the electropherogram is presented in Figure 1B. Both viruses contained
only 30-40S RNA of size class a. Five other recombinants
between the focus forming marker of a sarcoma and the host
range marker of a leukosis virus were studied and found to
contain only class a RNA. Some of these recombinants
showed the presence of class a and class b RNA initially;
however, subsequent cloning eliminated class b RNA. We
conclude that probably all recombinants carrying the focus
forming marker of a sarcoma and the host range marker of a
leukosis virus contain only 30-40S RNA of class a.

*Further evidence for the absence of class $\underline{b}$ RNA from
the 60-70S complex of sarcoma virus recombinants.*

Heated 60-70S RNA of the recombinants investigated
contains, in addition to a major component of 30-40S RNA,
minor heterogeneous RNA species of variable concentrations
(Figures 1-4). Therefore, it may be argued that these
minor RNAs are distinct subgenomic fragments, including
class b RNA of the parental leukosis virus perhaps acquired
by reassortment. To test this possibility, heat dissociated 60-70S ^{32}P RNA of PR-A x RAV-2 (Fig. 1A) was fractionated by sedimentation. Fractions comprising the 30-40S RNA
species and fractions comprising the minor heterogenous
species sedimenting at <30S and >10S were pooled separately
(Fig. 2A) and studied chemically. A sensitive method for
partial sequence comparison of RNAs has been developed by
Brownlee and Sanger (9) and is based on the electrophoretic
and chromatographic properties of RNase T1 resistant oligonucleotides. The technique has been used recently to compare tumor virus RNAs (8). It is shown in Figs. 2B and 2C
that the fingerprints of RNA pool 1 (>30S) and of RNA pool
2 (<30S and >10S) were indistinguishable. We conclude that
the minor heterogenous RNA species obtained after heat-
dissociation of 60-70S RNA consist predominantly of degraded 30-40S RNA of size class a rather than of chemically
distinct RNA species. However, the presence of low (<10%)
concentrations of small RNA species unrelated to 30-40S RNA
cannot be excluded by this experiment.

*Different recombinants derived from the same pair of
parental viruses and selected for the same markers have
RNAs of different size.*

Only two (host range and focus forming markers) of
presumably several genes, which may be exchanged between
leukosis and sarcoma viruses, have been selected for in the
recombinants studied here. If crossing over takes place

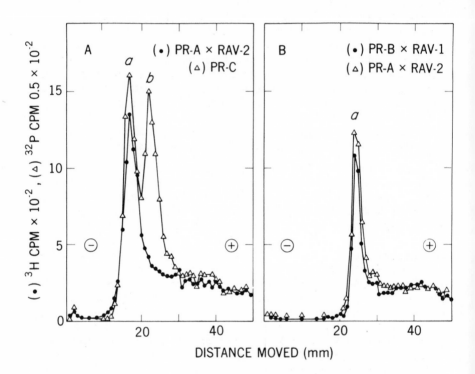

Fig. 1. Electrophoresis of heat-dissociated 60-70S
RNA of two cloned recombinant sarcoma viruses, PR-A x RAV-2
and PR-B x RAV-1. (A) Appropriate amounts of radio-labeled
PR-A x RAV-2 RNA and PR-C RNA were mixed and heated in
electrophoresis sample buffer and subjected to electrophoresis in 2% polyacrylamide as described (5). PR-C had
not been cloned recently and contained both class a and
class b RNA species. (B) A mixture of the RNAs of two
sarcoma virus recombinants PR-B x RAV-1 and PR-A x RAV-2
was analyzed as described for A.

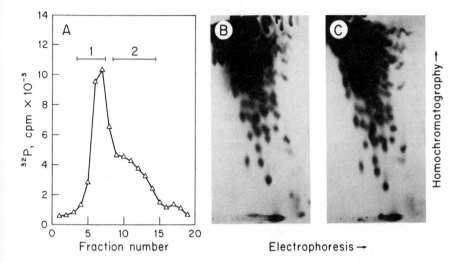

Fig. 2. Sedimentation and fingerprint analyses of heat-dissociated 60-70S ^{32}P-RNA (approx. 2 x 10^6 cpm) of a recombinant sarcoma virus PR-A x RAV-2, harvested at 3-5 hour intervals from infected cells. (A) RNA in 300 μl was heat-dissociated as described for Fig. 1. After addition of NaCl to 0.1 M the solution was layered on a 15-30% glycerol gradient containing 0.1   NaCl, 0.01   Tris HCl pH 7.4, 1 m   EDTA and 0.1% sodiumdodecylsulfate. Centrifugation was for 105 minutes at 50,000 rpm in a Spinco SW 50.1 rotor at 20°C. Fractions indicated by the bars in Fig. 2A were combined in two pools and the RNA was ethanol-precipitated. Fingerprinting of RNA pools 1 (B) and 2 (C) was as described previously (8).

between RNA tumor virus genomes, it may theoretically occur at any point on the genetic map between the focus forming and the host range markers. In this case, the RNAs of recombinants selected for the same markers, but derived from different cross over events could differ in their sequences. The first indication of such a difference was the observation that the 30-40S RNA of a recombinant between PR-B and RAV-3 (PR-B x RAV-3 #1) had a lower electrophoretic mobility and was therefore probably larger than the RNA of parental PR-B (Fig. 3A). The RNA of other preparations of this recombinant was also larger than class a RNA of PR-C and of another recombinant, PR-A x RAV-2 (Figs. 3C,E). By contrast class a RNAs of two different preparations of PR-B (Fig. 3B), class a RNAs of PR-B and PR-C (Fig. 3D), as well as class a RNA of PR-B and of PR-A x RAV-2 (Fig. 3F) were not distinguishable under our conditions. The RNAs of other recombinant clones between the focus forming marker of PR-B and the host range marker of RAV-3 fell into 3 electrophoretic classes: (i) PR-B x RAV-3 #2 had a lower mobility than parental class a RNA of PR-B (Fig. 4A). (ii) PR-B x RAV-3 #3 had a higher mobility than parental RNA (Fig. 4B). This recombinant was produced by transformed cells in 10-20 fold lower titers than other sarcoma viruses, perhaps indicating a defective replicating function. (iii) PR-B x RAV-3 #4 had practically the same mobility as parental PR-B RNA (Fig. 4C). These experiments indicate that the primary structure of recombinant RNAs differs from that of parental RNA.

The apparent molecular weight by which certain recombinant RNAs differ from parental, wild type RNA is estimated to be around 70,000 daltons on the following basis: The electrophoretic differences observed between RNAs were ± one fraction (Figs. 3,4). Class a and class b RNA differ by about 5 fractions under the same condition (*cf* Figs. 1, 3) (5,10). The difference between a and b was estimated to be about 350,000 daltons (10). Thus certain recombinant RNAs differ from wild type RNA by about one-fifth of that or 70,000 daltons. The size differences observed among the RNAs of distinct recombinants were stable after several successive clonings. This suggests that the size variations are not likely to be host modifications similar to those observed earlier in two specific cases which were not stable on passage of the virus in different cells (5).

The exact size of recombinant RNA cannot be predicted from the size of parental RNAs.

Class a RNAs of different nondefective sarcoma viral

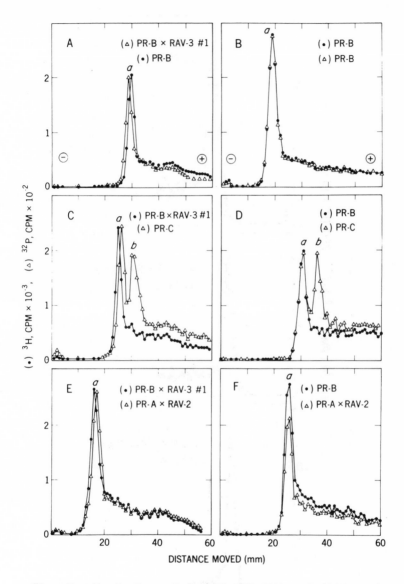

Fig. 3. Heat-dissociated 60–70S RNAs of different pre-parations of a cloned recombinant sarcoma virus, PR-B ×
RAV-3 #1 and of several other wild type and recombinant
sarcoma viruses after electrophoresis as described for
Fig. 1. The experiments were carried out to demonstrate
that class a RNA of PR-B × RAV-3 #1 had a lower electro-phoretic mobility than other class a RNAs as described in t
the text.

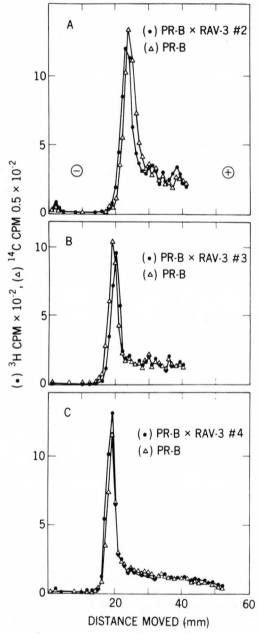

Fig. 4. The RNAs of three different recombinants PR-B × RAV-3 #2 (A), #3 (B) and #4 (C) after heat dissociation and electrophoresis with a standard of PR-B RNA. Conditions were as described for Fig. 1.

strains were shown to differ electrophoretically by about
one fraction (5); likewise class b RNAs of different
leukosis viruses differ slightly if compared by this
method (Fig. 5). Since the class a RNAs of PR-A, PR-B and
PR-C are all electrophoretically indistinguishable (Fig. 3,
ref. 5) but the class b RNAs of RAV-1, RAV-2 and RAV-3 used
to form recombinants with these sarcoma viruses are different, it appeared possible that a direct correlation existed
between the size of class b RNA in the parental leukosis
virus and the size of class a RNA in the recombinant virus.
For instance, it was found that compared to a class b RNA
standard (tdPR-C), the RNAs of RAV-1 and of RAV-2 are
slightly larger, while the RNA of RAV-3 is the same size as
the standard. Yet, the RNAs of two recombinants between
PR-B and RAV-3 (#1, Fig. 3 and #2, Fig. 4 respectively)
were actually larger than a recombinant between PR-A and
RAV-2 (cf. Figs. 1,3) and a recombinant between PR-B and
RAV-1 (cf. Fig. 1). Moreover, it was shown that class a
RNAs of different recombinant clones between the focus
forming marker of PR-B and the host range marker of RAV-3
have different mobilities. It follows that the exact size
of class a RNAs of different sarcoma virus recombinants
cannot be predicted from the known sizes of the parental
RNAs. This observation could be explained if some cross-
overs occurred at points of the parental genomes which were
not strictly homologous, leading to the acquisition or loss
of small stretches of genetic material in the recombinants
(unequal crossing over).

*Fingerprint-analyses of sarcoma virus recombinants
derived from the same pair of parental viruses and selected
for the same markers.*

If crossing over is responsible for the small electrophoretic differences observed among the class a RNAs of
four recombinants between PR-B and RAV-3, it would be expected that these RNAs also differ in their sequences.
This possibility was tested by fingerprinting the RNAs of
these four recombinants of the PR-B x RAV-3 cross (cf.
Figs. 3,4). It is apparent that their oligonucleotide
patterns are very similar but differ from each other in at
least 2-3 out of about 20 major RNase T1-resistant oligonucleotide spots (Fig. 6A-D). Some spots which are found
in one but not in all other recombinants are indicated by
arrows. The pattern of wild type PR-B is shown in Fig. 6E
and that of RAV-3 in Fig. 6F. Their patterns differ from
those of the recombinants more extensively than the recombinant patterns differ from each other.

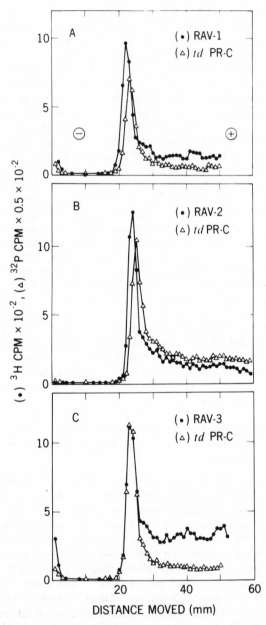

Fig. 5. The RNAs of three leukosis viruses RAV-1 (A), RAV-2 (B) and RAV-3 (C), used to form recombinants with PR RSV strains, after heat-dissociation and electrophoresis with a standard of *td* PR-C RNA. Conditions were as described for Fig. 1.

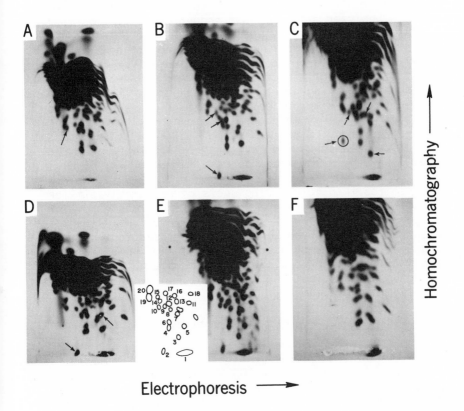

Fig. 6. Fingerprint analyses of the RNase T1-digested
60-70S ^{32}P-RNAs of the four recombinants PR-B × RAV-3 #1
(A), #2 (B), #3 (C) and #4 (D) as well as of PR-B (E) and
RAV-3 (F). 60-70S ^{32}P-RNAs of virus harvested at 12 hour
intervals were digested and analyzed as described previously (8) except that a 3% homo-mixture b (9) was used.
The arrows in A-D indicate spots not found in all of the
four recombinants analyzed. The circled spot in C has no
homologous counterpart in the patterns of either parental
virus (E,F). A schematic tracing of the large oligonucleotides of PR-B (E) identifies spots which were
analyzed as described in Table 1.

Although differences observed by fingerprinting are
only qualitative (8), it may be concluded that the recombinants analyzed differ in RNA sequences. This observation
supports the possibility that crossing over points between
focus forming and host range markers are not at a fixed
location. We have not determined in detail which of the
large oligonucleotides of the four PR-B x RAV-3 recombinants
are derived unchanged from either parental viral strain and
which of these oligonucleotides contain new sequences representing sites at which crossing over may have taken
place. However, at least one spot, circled in recombinant
#3, Fig. 6C, appeared to be new and not to have a homolo-
gous counterpart in either parental virus (Figs. 6E,F).
This spot as well as some others on Fig. 6A-D had lower
intensities than neighboring spots of presumably similar
size (9). This may be due to incomplete transfer of the
oligonucleotides from the cellulose acetate strip used for
electrophoresis to the DEAE thin layer used for homochroma-
tography (9). It may also reflect lack of homogeneities in
the RNAs. Further work, including complete transfer of
oligonucleotides, as used in Figs.6E,F will be required to
resolve this problem.

*The 60-70S tumor virus RNA appears to be largely
polyploid.*

If the 30-40S subunits of a given 60-70S tumor virus
RNA were identical and the 70S RNA represented a polyploid
genome, stable recombinants could arise only by crossing
over. The genetic complexity of the 60-70S RNA should then
be equal to that of each of the 30-40S pieces. However, if
60-70S RNA were haploid, its complexity would be higher
than that of an individual 30-40S subunit. The complexity
of an RNA species uniformly labeled with ^{32}P can be estimated if the sizes of several oligonucleotides derived from
it are determined, and the radioactivity of these oligonucleotides is compared with the total radioactivity in the
intact RNA molecule (11). The average complexity of PR-B
RNA as determined from about 20 RNase T1-resistant oligonucleotides, resolved as described in Fig. 6E, amounted to
2.7×10^6 daltons (Table 1). This is in good agreement
with the lower of several molecular weight estimates for
viral 30-40S subunits obtained by other methods (10).
Preliminary experiments suggest that the RNA of a recombinant has a similar complexity. Further work will be re-
quired to explain the fluctuations ($\sigma = \pm 0.48 \times 10^6$ daltons)
observed among complexity-estimates based on different oligonucleotides. These may be due to inhomogeneities of the

TABLE 1

THE COMPLEXITY OF PR-B RNA* ESTIMATED FROM THE SIZES
OF RNASE T_1-RESISTANT OLIGONUCLEOTIDES

Oligonucleotide spot no.[†]	CPM		Approximate base composition		Calculated complexity of RNA in daltons ($\times 10^{-6}$)	
	Exp.1	Exp.2	Exp.1	Exp.2	Exp.1	Exp.2
1	40,370	39,600	Poly A	Poly A	–	–[‡]
2	8,200	–	$(C_3A_7UG)_2$[§]	$(C_3A_6UG)_2$	3.2	2.9[†]
3	6,500	–	$C_3A_3U_4G$	$C_4A_4U_4G$	1.8	2.1
4	6,060	4,900	$C_6A_4U_3G$	$C_5A_3U_2G$	2.5	2.1
5	7,900	–	$(C_3A_2U_3G)_2$	$(C_3A_2U_4G)_2$	2.4	2.7
6	5,890	4,800	$C_5A_3U_3G$	$C_4A_3U_2G$	2.2	1.9
7	7,050	–	$(C_3A_3U_3G)_2$	$(C_3A_3U_3G)_2$	3.0	3.0
8	5,470	–	$C_4A_4U_3G$	$C_4A_5U_3G$	2.4	2.6
9	4,560	–	$C_4A_3U_3G$	$C_5A_4U_3G$	2.6	3.1
10	4,160	–	$C_5A_3U_2G$	$C_6A_4U_2G$	2.8	3.3
11	4,660	–	C_3AU_5G	C_3AU_5G	2.3	2.3
12	9,000	–	$(C_5A_3U_4G)_2$	$(C_5A_4U_3G)_2$	3.1	3.1
13	5,250	4,500	$C_2A_3U_3G$	$C_3A_4U_3G$	1.8	2.3
14	4,490	–	$C_4A_3U_2G$	$C_5A_4U_2G$	2.4	2.9
15	3,880	–	C_3A_3UG	$C_4A_4U_2G$	2.2	3.0
16	4,150	4,300	C_4AU_3G	–	2.3	–
17	5,940	5,000	$C_3A_4U_3G$	–	1.8	–
18	4,580	–	$C_2A_2U_4G$	$C_3A_2U_6G$	2.1	2.8
19	10,800	8,800	$(C_5A_4UG)_3$	$(C_4A_4UG)_3$	3.3	3.1
20	11,900	9,200	$(C_4A_4UG)_3$	$(C_4A_5UG)_3$	2.7	3.3
				Average[¶]	2.6	2.8

*60-70S ^{32}P-RNA derived from virus harvested at 12-hour intervals was prepared and exhaustively digested with RNase T_1 as described (8). The digest was resolved by electrophoresis and, after complete transfer to DEAE-cellulose, chromatographed as shown in Fig. 6. Two identical patterns were made each using 3.34 $\times$ 10^6 cpm (Exp. 1) or 2.86 $\times$ 10^6 cpm (Exp. 2) of the digested RNA. One pattern was used to determine the total radioactivity in a spot and the other to determine base compositions. An average of 380 cpm/nucleotide was found in Exp. 1. Oligonucleotides were eluted and base compositions determined by published procedures (9). Further details will be described elsewhere (Beemon and Duesberg, in preparation). The complexity was calculated using an average nucleotide MW of 323, calculated from the base composition of PR-B RNA (24.4% C, 23.8% A, 28.8% G, 23.0% U) and the known MW of the nucleotides.

[†]Numbers refer to diagram in Fig. 6E.

[‡]CPM from Exp. 1 and base compositions obtained in Exp. 2 were used in all calculations except where CPM from Exp. 2 are shown.

[§]Specific activity indicates more than one G per oligonucleotide due to either 2 (or 3) unresolved spots or to incompletely digested RNA. Heterogeneity of some spots is also suggested by their autoradiographic appearance; see for example spots #12, 19 and 20.

[¶]Values of presumed multiple spots are considered multiply in the average.

60-70S ^{32}P RNA prepared from virus harvested at 12 hr intervals (12,13,14). Oligonucleotides deriving from preferentially degraded sequences of 60-70S RNA would lead to a higher complexity-estimate and oligonucleotides derived from sequences of partially degraded RNA which associate preferentially with 60-70S RNA would lead to a lower complexity-estimate. Although these possible sources of error require further investigation we tentatively conclude that the RNase T1-resistant sequences of the 60-70S tumor virus RNA have an approximate complexity of 3×10^6. Thus, the 60-70S viral RNA appears largely polyploid and consequently recombination is likely to involve crossing over.

Discussion

Recombination involves crossing over.

Sarcoma virus recombinants, derived from a sarcoma virus parent with only class a RNA and a leukosis virus parent with only class b RNA, were found to contain only, or almost only, class a RNA. The class a RNA of certain recombinants was either larger or smaller than the parental class a RNA. Further, oligonucleotide patterns of recombinants, presumably derived from different recombination events but selected for the same markers, differed among themselves, and differed even more extensively from those of parental RNA. The 60-70S RNAs of wild type and of recombinant viruses appeared to be largely or completely polyploid. The sum of all these observations favors the conclusion that at least some of the recombination among avian tumor viruses involves crossing over.

If recombinants arose by reassortment of segments in a haploid genome, these recombinants should show only a limited number of fingerprint patterns. In recombinants selected for the same markers, the two segments containing these markers must be the same. Sequence diversity could still be caused by genome segments not carrying the selected markers, and of these there are at most two, to give a total of four segments per genome (4). The number of possible variations is then four; however, if there are only three segments per genome the same recombinants could occur in only two fingerprint variations. Since we have already observed four distinct fingerprint patterns in the PR-B x RAV-3 cross, our data would agree with reassortment only if the genome has four (but not three or two) genetically unique segments. Even in the case of four segments there is only a 9% chance for finding all four possible fingerprint patterns in the first four recombinants tested.

Disregarding our suggestive evidence on polyploidy, the remaining data on recombinant RNAs could be reconciled with reassortment, if we make the *ad hoc* assumption that class *b* RNA of leukosis virus is augmented by cellular RNA sequences to yield a class *a* molecule, when it becomes incorporated into a sarcoma virus in the process of recombination. This augmentation by cellular RNA would have to be genetically stable. Such a process could generate the observed size and sequence diversity and cannot be definitely ruled out on the basis of present data.

Is host-modification involved in the formation of recombinant RNA?

Small differences in the size of class *a* RNA can be observed among recombinants selected for the same markers. These differences could result from host modification of the viral RNA rather than from unequal crossing over. Such modifications may include various degrees of polyadenylation (15) or addition of cellular sequences acquired at the chromosomal sites at which viral DNA is thought to integrate into cellular DNA. Although such changes could account for the electrophoretic differences observed among different recombinant RNAs, it is unlikely that these variations in size amounting to only ±2% of the RNA are also responsible for the changes observed in fingerprints patterns.

Alternatively recombination between exogenous tumor viruses could include interactions with endogenous tumor viruses present in all normal chicken cells (16). This type of "host-modification" has not been tested for by our experiments. However, recombination with endogenous virus has been observed only in helper factor positive cells, in which endogenous virus is at least partially expressed (3) but not in the helper factor negative cells used to prepare our recombinants. Also, if genetic interactions with an endogenous virus were responsible for some of the new properties of recombinant RNAs, it would be difficult to explain why such modification of RNA is not regularly observed in single infection.

What is the mechanism of tumor virus recombination?

No direct answer can be given to this question from our experiments, except that crossing over appears to occur. Since there is no precedent and no plausible molecular mechanism for high frequency crossing over between viruses containing single-stranded RNA, it appears likely that recombination among avian RNA tumor viruses involves the synthesis of the DNA provirus (17,18). The high frequency recombination among RNA tumor viruses could then be a direct consequence of polyploidy. The progeny of a doubly infected cell would be largely heterozygous, containing different genomes in a 60-70S complex. Transcription of such a heterozygous RNA into DNA would bring homologous DNAs together and could increase the chances of crossing over (3,6,17).

List of Virus Abbreviations

PR-A: Prague Rous sarcoma virus, subgroup A.
PR-B: Prague Rous sarcoma virus, subgroup B.
PR-C: Prague Rous sarcoma virus, subgroup C.
*td*PR-C: Transformation defective derivative of PR-C.
RAV-1:   Rous associated virus, type 1, subgroup A.
RAV-2:   Rous associated virus, type 2, subgroup B.
RAV-3:   Rous associated virus, type 3, subgroup A.

Acknowledgments

We thank Sunny Kim, Marie Stanley, Philip Harris, Annie Chyung, Hsiao-Ching Pang and Joseph Green for excellent help with these experiments. The work was supported by Public Health Service research grants CA 11426 and CA 13213 from the National Cancer Institute and by the Virus Cancer Program-National Cancer Institute contracts No. NO1 CP 43213 and No. NO1 CP 43242.

The same data have been presented in "Mechanisms of Virus Disease; ICN-UCLA Symposia in Molecular and Cellular Biology", Vol. I, 1974 (ed., W. A. Benjamin Inc., Menlo Park, California).

References

1. P. K. Vogt, *Virology* 46,947,(1971).
2. S. Kawai and H. Hanafusa, *Virology* 49,37,(1972).
3. R. A. Weiss, W. Mason, and P. K. Vogt, *Virology 52*, 535,(1973).
4. P.H. Duesberg, "Current Topics in Microbiology and Immunology" *51*,79,(1970).
5. P.H. Duesberg and P.K. Vogt, *Virology 54*,207,(1973).
6. P.K. Vogt and P.H. Duesberg, "Virus Research", p.505, (1973).
7. G.S. Martin and P.H. Duesberg, *Virology 47*,494,(1971).
8. M.M. Lai, P.H. Duesberg, J. Horst, and P. K. Vogt, *Proc. Nat. Acad. Sci. 70*,2266,(1973).
9. G.G. Brownlee, and F. Sanger, *Europ. J. Biochem. 11*, 393,(1969).
10. P.H. Duesberg and P.K. Vogt, *J. Virol. 12*,594,(1973).
11. W. Fiers, L. Lepoutre and L. Vandendriesche, *J. Mol. Biol. 13*,432,(1965).
12. P.H. Duesberg and B. Cardiff, *Virology 36*,696,(1968).
13. J.P. Bader and T.C. Steck, *J. Virol. 4*,454,(1969).
14. R.L. Erikson, *Virology 37*,124,(1969).
15. M.M. Lai and P.H. Duesberg, *Nature 235*,383,(1972).
16. R.A. Weiss, "RNA Viruses and Host Genome in Oncogenesis", p.117,(1971).
17. P.K. Vogt, "Proceedings of the Fourth Lepetit Colloquium", p.35,(1973).
18. R.A. Weiss, "Proceedings of the Fourth Lepetit Colloquium", p.130,(1973).

ON THE ORIGIN OF RNA TUMOR VIRUSES

Howard M. Temin
McArdle Laboratory, University of Wisconsin,
Madison, Wisconsin 53706

A good deal of recent evidence is consistent with the hypothesis (protovirus hypothesis) that ribodeoxyviruses (viruses whose virions contain RNA and a DNA polymerase) arose from normal cellular components (Temin, 1974b). This evidence is especially strong for the two groups of avian ribodeoxyviruses -- the avian leukosis-sarcoma viruses (ALV) and the reticuloendotheliosis viruses (REV).

The ALV and REV virions are similar in having C-type morphology and containing 60-70S RNA and a DNA polymerase (see Temin, 1974a). Both ALV and REV replicate through a DNA intermediate, the DNA provirus, as shown by nucleic acid hybridization and infectious DNA experiments. They also have a requirement for activation of virus production by a replicative cell cycle.

ALV and REV virions differ in the presence or absence of endogenous RNA-directed DNA polymerase activity -- ALV virions have this activity, REV virions do not; in neutralization of infectivity and of DNA polymerase activity; in the sequences of the virion 60-70S RNA; and in their pathology in fowl and their effects on cells in culture.

To look for relationships between ALV, REV, and avian cells, studies of nucleic acid sequence homology and of serological relationships of DNA polymerases were performed. RNA from Rous-associated virus-0 (RAV-0), an ALV, hybridized 70% to DNA of uninfected chicken cells, 15% to DNA of ring-necked pheasant cells, 5% to DNA of quail cells, and 0% to DNA of duck cells (Neiman, 1972; 1973; Kang and Temin, 1974). RNA from Trager duck spleen necrosis virus, an REV,

———————
Summary of a talk presented April 2, 1974 at the 1974 Flexner Symposium, Vanderbilt University. A fuller version of this talk will be published in the Harvey Lectures, 1973-1974. The research in my laboratory was supported by U.S. Public Health Service research grant CA-07175 from the National Cancer Institute and grant VC-7 from the American Cancer Society. I hold Research Career Development Award K3-CA-8182 from the National Cancer Institute.

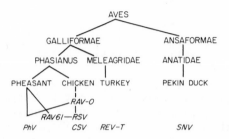

Fig. 1. Possible evolutionary relationships of avian ribodeoxyviruses and fowl. Aves is the class; Galliformae and
Ansaformae are orders; Phasianus, Meleagridae, and Anatidae
are families. RAV-0, Rous-associated virus-0; RSV, Rous
sarcoma virus; and RAV-61, Rous-associated virus-61 are
avian leukosis-sarcoma viruses. PhV, pheasant virus; CSV,
chick syncytial virus; REV-T, reticuloendotheliosis virus
(strain T); and SNV, spleen necrosis virus are reticuloendotheliosis viruses.

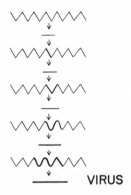

Fig. 2. A possible mechanism for the origin of ribodeoxyviruses. A section of a cell genome becomes modified in
successive DNA (∿∿) to RNA (-) to DNA transfers until it
becomes a ribodeoxyvirus genome. It evolves as part of the
cell genome until it becomes a virus genome. Then it
evolves independently. The time scale may be millions of
years in germ-line cells and days in somatic cells.

hybridized 10% to DNA of uninfected chicken, quail, turkey, and pheasant cells, and 0% to DNA of duck cells (Kang and Temin, 1974).

The DNA related to RAV-0 in uninfected chicken cells does not represent the complete provirus of an infectious RAV-0. No infectious DNA for RAV-0 was isolated from chicken cells not producing RAV-0 (Cooper and Temin, 1974).

The DNA polymerases of ALV, REV, and avian cells are serologically related (Mizutani and Temin, 1974). Antibody to the chicken large DNA polymerase neutralized and was blocked by REV DNA polymerase. Antibody to avian myeloblastosis virus (an ALV) DNA polymerase neutralized and was blocked by REV DNA polymerases and was blocked by chicken large and small DNA polymerases. Antibody to Trager duck spleen necrosis virus (an REV) DNA polymerase neutralized and was blocked by chicken large and ALV DNA polymerases and was blocked by chicken small DNA polymerase.

These relationships among ALV, REV, and avian cell DNA polymerases and nucleic acid sequences are like those among the different species of fowl. They indicate that ALV and REV can be put in an evolutionary tree of fowl relationships (Figure 1).

The ribodeoxyviruses might have evolved from a portion of the cell genome by successive DNA to RNA to DNA information transfers as indicated in Figure 2.

References

1. G. M. Cooper and H. M. Temin, in preparation (1974).
2. C.-Y. Kang and H. M. Temin, in preparation (1974).
3. S. Mizutani and H. M. Temin, J. Virology 13, 1020 (1974).
4. P. Neiman, Science 178, 750 (1972).
5. P. Neiman, Virology 53, 196 (1973).
6. H. M. Temin, Advan. Cancer Res. 19, 47 (1974a).
7. H. M. Temin, Annu. Rev. Genetics 8, in press (1974b).

ENDOGENOUS LEUKOSIS VIRUSES OF PHEASANTS

Donald J. Fujita, Young C. Chen, Robert R. Friis
and Peter K. Vogt

Department of Microbiology, University of Southern California
School of Medicine, Los Angeles, California

Summary

*Endogenous leukosis-like viruses have been isolated
from normal cells of ring-necked and golden pheasant embryos.
All isolates from ring-necked pheasants are related to each
other and belong to subgroup F. The endogenous viruses obtained from golden pheasants are also related to each other
but differ in their envelope properties from all known avian
RNA tumor viruses; they are assigned to subgroup G. Subgroups
F and G viruses can infect most avian cell types, recombine
with nondefective avian sarcoma viruses, and provide helper
activity for the defective Bryan high titer strain of Rous
sarcoma virus.*

Introduction

Recently T. and H. Hanafusa (1973) described the isolation of a leukosis type virus from normal cells of ring-
necked pheasants (*Phasianus colchicus torquatus*) and assigned
it to a new avian tumor virus subgroup, F. We have confirmed
the widespread occurrence of subgroup F virus in ring-necked
pheasant embryos and have also isolated an endogenous leukosis like virus from golden pheasant embryos (*Chrysolophus
pictus*). The present report summarizes the properties of
ring-necked pheasant and of golden pheasant viruses. A
detailed account of these studies will be published elsewhere
(Fujita *et al.*, 1974).

Results

Endogenous helper activity in normal pheasant cells.

The defective Bryan high titer strain of RSV (BH RSV)
requires the cooperation of a helper virus to produce progeny with a functional envelope. Endogenous leukosis viruses
can supply this helper function (Weiss, 1969; Hanafusa *et al.*,
1970; Hanafusa and Hanafusa, 1973). It is therefore possible
to use BH RSV as a tool for the detection of endogenous RNA
tumor viruses. If the endogenous virus complements RSV,
infectious sarcoma viral pseudotype will be produced.

We have used quail cells transformed by BH RSV in the absence of a helper to search for endogenous leukosis virus in normal pheasant cells. These R(-) quail cells were fused by cocultivation or with the aid of inactivated Sendai virus with normal fibroblasts of ring-necked or golden pheasant origin. Virus production in the mixed cultures was followed for a period of 6 to 8 weeks. Tables 1 and 2 show the results of these experiments. Fibroblasts from 6 out of 8 golden pheasant embryos provided helper activity, and among ring-necked pheasant embryos 4 out of 5 were positive within the period of observation. The RSV pseudotypes from pheasant cultures plated on C/E cells; therefore they were different from the subgroup E virus produced with the endogenous helper factor found in chicken cells. In contrast to the situation in the chicken, where there is a good correlation between helper activity and presence of group specific viral antigens in normal cells (Weiss, 1969; Hanafusa *et al.*, 1970; Chen and Hanafusa, 1974; Vogt *et al.*, 1973), neither species of pheasant showed a link between presence of *gs* antigen and helper factor.

The isolation of infectious RSV pseudotypes by cocultivation of R(-) quail cells with pheasant cells can be criticized on two counts. One, it is conceivable that the fusion activates an endogenous virus in quail rather than in pheasant cells; and two, a helper virus could be introduced in the system with the inactivated Sendai virus which had been grown in chicken cells. In order to rule out these possibilities foci produced on ring-necked pheasant cells by RSV(RAV-0) (a subgroup E pseudotype of BH RSV) at very low multiplicities were picked, cultivated individually, and tested for the appearance of infectious RSV pseudotype which was not subgroup E. Table 3 shows that 3 out of 7 single foci produced large amounts of such a virus; two released low titers, one was nonproductive, and one produced subgroup E virus; this latter was probably doubly infected with RSV and RAV-0. The synthesis of infectious pseudotypes by pheasant cells alone after single infection with BH RSV rules out a contribution of the quail cells or of the Sendai virus to the helper activities summarized in Tables 1 and 2.

We will refer to the RSV pseudotype produced with helper activity from ring-necked pheasant cells as RSV(RPV), and the pseudotype from golden pheasant cells will be termed RSV(GPV).

Separation of Helper Viruses From RSV(RPV) and RSV(GPV).

The helper viruses presumed to be present in the pseudotype stocks from pheasant cells were isolated by the endpoint interference technique (Rubin and Vogt, 1962). Tables 4 and

TABLE 1

INFECTIOUS RSV AFTER FUSION OF R(-) QUAIL

WITH NORMAL GOLDEN PHEASANT CELLS

Golden pheasant	Titer (FFU/ml) of RSV on C/E		
embryo number	12	23	35 days
1 $gs-$	0	0	NT
2 $gs-$	0	5	10
3 $gs-$	50	3×10^3	8×10^3
4 $gs-$	0	0	1×10^2
5 $gs-$	10	5	1×10^2
6 $gs-$	0	0	NT
7 $gs-$	0	0	50
8 $gs-$	5	4×10^2	5×10^3

TABLE 2

INFECTIOUS RSV AFTER FUSION OF R(-) QUAIL

WITH NORMAL RING-NECKED PHEASANT CELLS

Ring-necked pheasant	Titer (FFU/ml) of RSV on C/E		
embryo number	23	35	52 days
1 $gs-$	0	0	3×10^3
2 $gs-$	5	30	8×10^2
3 $gs-$	0	0	NT
4 $gs+$	0	5	0
5 $gs+$	0	10	NT

TABLE 3

SINGLE FOCI OF RSV(RAV-O) ON RING-NECKED PHEASANTS:
PRODUCTION OF A NEW PSEUDOTYPE

Focus	RSV titer (FFU/ml)		
number	Japanese quail	Chicken C/BE	Chicken C/E
1	1.7×10^3	1.6×10^3	1.7×10^3
2	8.0×10^3	5.5×10^3	6.0×10^3
3	1.3×10^2	1.7×10^2	1.3×10^2
4	10	30	>10
5	40	10	20
6	0	0	0
7	$>10^4$	0	10

TABLE 4

ISOLATION OF RING-NECKED PHEASANT VIRUS FROM RSV PSEUDOTYPE

Dilution of	Foci formed by challenge virus		
pseudotype	RSV(RPV)	RSV(GPV)	No challenge
1:3,200	17	83	14
1:6,400	0	182	0
1:12,800	366	60	0
1:25,600	296	89	0
uninfected	474	77	0

5 summarize these experiments and demonstrate interference with homologous sarcoma virus challenge beyond the endpoint of transformation. From these resistant cultures the leukosis viruses RPV and GPV were obtained.

Host Range of RSV(RPV) and RSV(GPV).

Table 6 summarizes the plating efficiencies of RSV(RPV) and RSV(GPV) on various avian cell types. Both viruses can infect cells which are genetically resistant to avian tumor viruses of subgroups A through E, but both are excluded from goose cells. RSV(RPV) plates well on duck cells, but RSV(GPV) does not. We conclude that the RSV pseudotypes with endogenous pheasant viruses do not belong to subgroups A through E, and that GPV constitutes a different subgroup from RPV. Since RPV was assigned to subgroup F by T. and H. Hanafusa (1973), we propose to classify GPV as subgroup G.

Interference Patterns of Subgroups F and G.

The conclusions derived from host range studies were supported by observations on viral interference (Table 7). Cells infected with subgroups A to E leukosis viruses were not refractory to RSV(RPV) or RSV(GPV), while showing strong resistance against homologous challenge viruses. RAV-61, kindly provided by Dr. H. Hanafusa as a subgroup F standard, interfered with RSV(RPV) as did RPV, but not with RSV(GPV). GPV interfered only with RSV(GPV). In some experiments RPV also caused moderate resistance to a subgroup A challenge (EOP 0.05 to 0.1), but this was not observed consistently. On quail fibroblasts RPV showed heterologous interference with subgroup E challenge viruses; this observation has also been made by T. and H. Hanafusa (1973).

Serological Properties of Subgroups F and G.

As expected, antisera against subgroups A to D failed to neutralize RSV(RPV) or RSV(GPV) (Table 8). There was also no correlation between subgroup E neutralizing activity of a serum and neutralization of subgroup F (Table 9). Some subgroup E sera also neutralized RSV(RPV), but so did a small proportion of normal ring-necked pheasant sera, indicating that immunization with subgroup E was not the cause of the anti-F activity. The fact that some normal pheasant sera contained antibody against the endogenous virus suggests that pheasants are not immunologically tolerant to this subgroup; since they are fully susceptible to exogenous infection by subgroup F virus, horizontal spread resulting in an immune

TABLE 5

ISOLATION OF GOLDEN PHEASANT VIRUS FROM RSV PSEUDOTYPE

Dilution of	Foci formed by challenge virus		
pseudotype	PR-A	RSV(GPV)	no challenge
1:100	30	34	2
1:300	45	1	0
1:900	62	356	0
1:1,800	70	400	0
uninfected	96	488	0

TABLE 6

HOST RANGE OF SUBGROUPS F AND G

	Virus	
Cell type	RSV(GPV)	RSV(RPV)
Chicken C/E	S^a	S
C/A	S	S
C/BE^b	S	S
C/ABE	S	S
C/C	S	S
Japanese quail	S	S
Ring-necked pheasant	S	S
Golden pheasant	SR	S
Peking duck	R	S
Goose	R	R

^{a} S = EOP 0.25 to 1.0 relative to C/E
SR = EOP 0.01 to 0.25
R = EOP <0.01

^{b}Avian cells resistant to subgroup B also show strongly reduced susceptibility for subgroup D.

TABLE 7

INTERFERENCE PATTERNS OF SUBGROUPS F AND G

Cell	Interfering virus and subgroup		Challenge virus RSV(RPV)	Challenge virus RSV(GPV)
C/E	none		S^a	S
	RAV-1	(A)	S	S
	RAV-2	(B)	S	S
	tdB77	(C)	S	S
	CZAV	(D)	S	S
	RPV	(F)	R	S
	GPV	(G)	S	R
	RAV-61	(F)	R	S
Ring-necked pheasant	none		S	S
	RAV-0	(E)	S	S

^{a}See footnote Table 6

TABLE 8

ANTISERA AGAINST SUBGROUPS A TO D FAIL TO NEUTRALIZE

RSV(RPV) OR RSV(GPV)

Serum	Fraction surviving FFU Subgroup and virus					
	A RSV(RAV-1)	B RSV(RAV-2)	C PR-C	D SR-D	F RSV(RPV)	G RSV(GPV)
Control	1.00	1.00	1.00	1.00	1.00	1.00
Anti-A	<0.01				0.89	1.17
Anti-AB	0.02	0.10			0.89	0.94
Anti-C			0.02		0.77	1.01
Anti-D				0.10	0.70	1.62

TABLE 9

SEROLOGICAL DIFFERENTIATION BETWEEN SUBGROUPS E, F, AND G

Serum source	Prepared against	Serum number	Percent surviving FFU		
			RSV(GPV)	RSV(RPV)	RSV(RAV-0)
Ring-necked pheasant	E	887	74	3	<0.2
	normal	634	126	12	102.0
Rabbit	E	1	67	109	<0.001

response could occur. Subgroup F neutralizing sera did not affect subgroup G, and no sera active against subgroup G have been found.

Defectiveness of RSV(RPV) and RSV(GPV).

The interaction of BH RSV with its helper virus is a special case of phenotypic mixing (Hanafusa *et al.*, 1963); and BH RSV remains genetically defective, probably because it is unable to undergo recombination with the helper virus in the genes controlling host range (Kawai and Hanafusa, 1972; Weiss *et al.*, 1973). The pseudotypes of BH RSV and endogenous pheasant viruses conform to this rule (Table 10). In an infectious center assay which makes the formation of foci strictly dependent on the release of infectious virus rather than the division of the initially infected cell (Weiss *et al.*, 1973), RSV(RPV) and RSV(GPV) failed to register; whereas they did form foci if experimental conditions permitted cell division. A nondefective control virus registered equally well in the direct focus assay and in the infectious center test.

Recombinants Between RPV and Prague Strain Rous Sarcoma Virus.

Prague strain Rous sarcoma virus (subgroup B) was grown together with RSV(RPV), and from the mixed harvests clones of sarcoma virus were isolated which could form infectious centers on C/BE cells. This property is not present in either parent: RSV(RPV) fails to form infectious centers (Table 10), although it can infect C/BE cells, and PR-B cannot enter C/BE cells but is nondefective and registers in the infectious center assay. The recombinant selected thus combines the host range of RPV with the productive focus forming ability of PR-B. Table 11 shows the behavior of one such recombinant clone in the infectious center assay: it no longer has a subgroup B host range but is still nondefective. RPV can therefore donate its host range marker to PR-B in genetic recombination. Preliminary experiments indicate that GPV can also combine with nondefective RSV.

Discussion

RPV and GPV were found in the majority of pheasant embryos tested. Although the number of embryos investigated is still small, the available data suggest that these endogenous viruses are of widespread occurrence; thus RPV was isolated from East Coast pheasant flocks by T. and H.

TABLE 10

DEFECTIVENESS OF RSV(GPV) AND RSV(RPV)

Virus	Number of FFU	Number of Infectious centers
PR-A	85	80
RSV(GPV)	446	1
RSV(RPV)	358	0

TABLE 11

NONDEFECTIVENESS OF RECOMBINANTS

BETWEEN PR-B AND RPV

Virus	Number of FFU on C/E cells	Number of infectious centers on C/BE cells
PR-B	201	0
PR-A	336	277
PR:RPV#7	300	366

Hanafusa, and we have seen it in material from California and from the State of Washington.

RPV causes lymphoid leukosis in chickens (Purchase, private communication), but its oncogenic potential in the host species of origin remains to be tested. The possible leukemogenic action of GPV is also still under investigation.

The RPV and GPV derived from RSV pseudotypes grow to high titers in several types of avian cells. In contrast to these pheasant helper viruses isolated from BH RSV pseudotypes an endogenous virus found by T. and H. Hanafusa in ring-necked pheasant cells without the aid of BH RSV grew extremely poorly and retained very low titers in continuous culture (T. and H. Hanafusa, 1973). It is therefore possible that only part of the RPV or GPV genomes are endogenous to pheasant cells, and that the viral isolates from BH RSV pseudotypes have acquired some genetic information from the sarcoma virus. Indeed, Kang and Temin (1973) have demonstrated that RAV-61, which appears very similar, if not identical, to RPV shows close nucleic acid homology with RAV-0. Yet Neiman (1973), using a RAV-0 probe, was able to find only part of its sequences contained in the genome of normal *gs+* or *gs-* pheasant cells. It will now be of interest to use nucleic acid probes of RPV and of GPV to determine the extent to which these viral genomes are represented in normal ring-necked and golden pheasant cells.

Acknowledgments

This study was supported by U. S. Public Health Service Grant No. CA 13213 and Contract No. N01-CP-43242 from the National Cancer Institute. D. Fujita received a fellowship from the Leukemia Society of America.

List of Abbreviations

FFU: Focus forming units
RSV: Rous sarcoma virus
BH: Bryan high titer strain
R(-):   Defective BH RSV without helper
gs: Group specific antigen
C/E, C/BE, C/A, C/ABE, C/C: Chicken cell types listing the excluded viral subgroup after the bar
RPV: Ring-necked pheasant virus
GPV: Golden pheasant virus
PR-A: Prague RSV subgroup A
PR-B: Prague RSV subgroup B
RAV: Rous associated leukosis virus

tdB77: Transformation defective variant of avian sarcoma virus B77
CZAV: Carr Zilber associated virus; a leukosis virus isolated from Carr Zilber RSV
SR-D: Schmidt Ruppin RSV of subgroup D
EOP: Efficiency of plating

References

1. J. H. Chen, and H. Hanafusa, *J. Virol.13*,340,(1974).
2. T. Hanafusa and H. Hanafusa, *Virology 51*,247,(1973).
3. H. Hanafusa, T. Hanafusa, and H. Rubin, *Proc. Nat. Acad. Sci. U.S.49*,572,(1963).
4. T. Hanafusa, T. Miyamoto, and H. Hanafusa, *Proc. Nat. Acad. Sci.U.S.66*,314, (1970).
5. C. Kang and H. Temin, *J. Virol.12*,1314,(1973).
6. P. Neiman,*Virology 53*,196,(1973).
7. H. Rubin and P. Vogt, *Virology 17*,184,(1962).
8. P. Vogt, R. Friis, and R. Weiss, *Proc. 7th Natl. Cancer Conference*, p.55,(1973).
9. R. Weiss, *J. gen. Virol.5*,511,(1969).
10. R. Weiss, W. Mason, and P. Vogt, *Virology 52*,535,(1973).
11. D. Fujita, Y. Chen, R. Friis, and P. Vogt, *submitted to Virology*,(1974).

A 4
B 5
C 6
D 7
E 8
F 9
G 0
H 1
I 2
J 3